Developing Expert CAD Systems

W0235031

Developing Expert CAD Systems

Vivienne Begg

**Kogan
Page**

This first edition published by Kogan Page Ltd
120 Pentonville Road, London N1 9JN
Reprinted 1987

Copyright © 1984 Kogan Page
All rights reserved

British Library Cataloguing in Publication Data

Begg, Vivienne
    Developing Expert CAD systems.
    1. Engineering design — Data processing
    I. Title
    620'.00425'02854   TA174

ISBN-13: 978-1-85091-298-9   e-ISBN-13: 978-94-009-3405-4
DOI: 10.1007/978-94-009-3405-4

Printed in Great Britain by Billing & Sons Limited,
Worcester

Contents

Chapter 1: Introduction   7

The importance of CAD to electronics technology, 7
The complexity of modern electronics design, 9

Chapter 2: What *is* CAD?   12

The use of CAD, 13
Problems arising from the use of CAD, 13
The use of CAD in electronics design, 16
Conclusions, 17

Chapter 3: How is design carried out?   19

The design process, 19
Circuit analysis, 23
Logic design, 24
Logic simulation, 24
   Logic to layout, 24
Manual layout procedures, 26
Automatic layout routines − PCBs, 27
   Preliminaries, 27
   Packaging, 28
   Automatic placement routines, 28
   Automatic routing, 28
Automatic layout routines − ICs, 29
Design rule checkers, 30
Mask making, 31
Conclusions, 32

Chapter 4: What makes a good CAD system?   33

Communication through the human-computer interface, 33
The CAD users' needs, 35
   The user model, 38
   The user profile, 38
Schemes of communication, 43

Chapter 5: Design representation   45

Problem formalization, 45
Abstraction, decomposition and refinement, 49
Concepts of formal and informal knowledge, 57

Chapter 6: Techniques for electronics engineering
from artificial intelligence   63

The evolution of design methods, 64
Formal and informal decision-making procedures, 67
The problem-solving approach, 70
Design automation, 73
Building intelligence into design procedures, 76
   The computer as consultant, 77
   Choosing languages for design, 79
   Guiding search through a design library, 80

Chapter 7: An 'ideal' CAD system   85

Components of a system, 85
Basic structure and operation, 87
A sketch of a computer aid for specification, 89
The CONSULTANT interface, 92
Advantages of an intelligent CAD system, 95

Chapter 8: A summary   97

Glossary, 103

References, 110

Index, 115

CHAPTER 1

Introduction

The importance of CAD to electronics technology
Computer-aided design (CAD) is one way of coping with the
problem of how to design and build very complex systems.
This problem is particularly acute in electronics technology.
Designs are now (1984) said to be design-limited, rather than
technology-limited. It can take months to generate a design
for a chip, so that it might be obsolete before it can be
manufactured. Manual design of large-scale integration (LSI)
chips *(circa* 10,000 gates) is almost impossible. However,
using current technology it is possible to produce chips
having 250,000 gates. It is understandable, therefore, that
there is great interest in improving existing CAD systems.

Designers of CAD systems are concerned with formalizing
and automating as much of the design task as possible.
Automating design of any kind has long been acknowledged
as a project fraught with intractable problems. A human
designer has to have an understanding of the nature of the
materials used in manufacture, a knowledge of common
problems and well-tried solutions, and above all, creativity in
producing new designs. Understanding, knowledge and
creativity are three properties even the most artificially
intelligent of computer programs have been entirely lacking
in until very recently. Some people would deny computers
these qualities entirely, on philosophical grounds (eg Dreyfus
1979; Searle 1981). There are few theories in cognitive
psychology which can help. Understanding and knowledge
are very much the territory of the epistemological philosopher,
and creativity has long been thought of as an almost magical
attribute of some gifted humans.

The magical quality possessed by the mythical designer
often obscures the rather prosaic methodologies used by

designers in electronics engineering in their everyday work. An electronics designer does not share the limelight which is the due of other industrial designers, such as those involved in creating new designs for cars, furniture or clothes. High technology is an obscure field for non-technologists: the idea of a 'fashionable' new chip is not viable, because however marvellous the achievement in engineering terms, and however fine the aesthetics of the product (electronics engineers *do* have a keen sense of the beautiful) the 'nuts and bolts' of a chip are invisible and intangible. Modern electronics engineering is more arcane than alchemy. The production of semiconductors is carried out in better than surgically clean conditions, on mostly automated production lines tended by highly trained, priest-like operatives. Design is also being automated by design teams whose relationship with their computer aids can be equally exclusive.

The paradox is that the more automated a system becomes, the better understood it has to be, but that understanding only has to be possessed by the programmer and not, as in unaided design, by a caste of craftsmen. Designers using CAD for the first time often feel that their expertise will be eroded and nullified by the computer. They are often delighted to discover that the computer erodes only the aspects of their job that were stultifyingly boring, and gives them the opportunity of acquiring new skills, not less satisfying than the old skills, but transcending them. Design methodology is being investigated and simulated in the process of creating design tools. Looking at these design tools and discovering how they are used is one way of finding out what design is. As a first approximation, we can assert that design is whatever the designer does in order to produce a plan of the required electronic component or system from a specification. Some typical CAD systems, and some of the problems they pose for users, are described in Chapter 2.

For years, designers have carried out their job successfully without computer aids. In the history of engineering, the concept of designer is a fairly new one. Practical, domestic machines, such as houses, carts, mills, clocks, locks and so on, were for centuries produced by guilds of craftsmen for whom design meant changes in surface decoration. Everyone knew what a house or a clock looked like, so no one had to draw plans or formulate explicit descriptions of how structures were put together: changes in basic structural design came very slowly, by evolution. The industrial revolution had some impact on this approach to engineering design, for two reasons: first, growth in manufacturing industry created a need for more engineers, and second, the beginnings of automation required that skills were analysed and made explicit. What used to be learned only after years of apprenticeship as a craft skill had to be assimilated in a much shorter course of technical education as engineering principles. For the purposes of mass production, the range of skills possessed by a single craftsman were broken down into their component parts, simple actions that could be performed by a less skilled person or by a machine.

Nevertheless, these changes did not excessively accelerate the pace of change in industrial design. For example, the motor car has scarcely undergone one major technological development since it was invented. Body shapes change, engine performance is refined and improved, but the structure of an engine and the configuration of wheels, chassis, body, etc are directly analogous in a Model T Ford and a 1984 model.

The complexity of modern electronics design

In electronics engineering (specifically, digital electronics) technological change has been very rapid, especially since the mid-1960s. Materials have changed: circuits were implemented with what seem now to be massive valves, 20 years ago germanium transistors soldered on to boards were used. Now, the same functions can be performed by a die cut from a slice of silicon crystal (a wafer) which has been etched by exotic acids, baked with poisonous gases at high temperatures and bombarded with ions. All the techniques used involve rare, dangerous, delicate and comparatively poorly understood materials and processes. There is less correspondence between the technology of the first computers of the 1940s and 50s and those of today than there is between that of the Model T Ford and that of the space shuttle.

Apart from the bewildering changes in implementation technology over less than one working lifetime, there have been changes in application which have led to massive changes in design methodology. Computing and control applications have become wide ranging as the technology allows the implementation to be physically smaller and smaller. The corresponding increase in complexity of systems being designed has forced designers to work in teams, and to attempt to automate certain aspects of the design task, in the form of CAD tools. Teamwork necessitates good communication. It rapidly became obvious to users of early CAD tools that unless computer output was easy to understand and human input to computers was carefully phrased, the systems didn't work. The savings in time made by automating the boring jobs were eroded by the time spent translating inputs into 'computerese' and poring over manuals. Chapter 4 suggests a set of criteria which define a good human-computer interface for a CAD system in terms of what is available at present and what it will be useful to have in the future.

Apart from rationalizing the human-computer interface, there are other, more radical issues to be tackled in improving CAD systems. The main problem facing electronics designers is system complexity. There are many different sorts of measures of complexity, but for these purposes the following definition will suffice: a complex object is one with sufficiently many interconnected parts that a human observer, given an understanding of the materials from which it is made, cannot interpret its structure well enough to predict its behaviour.

Electronics systems have become too complex to be designed by one person: during 1976-77 it became obvious to most people in semiconductor design that LSI circuits were becoming too complex to be designed using the current design methodology. That methodology was inherited from the days of printed circuit boards (PCBs) and is usually referred to as *random logic*. One man worked on a chip design from beginning to end, holding most of its design in his head, in a manner similar to that known to the software engineering fraternity as 'hacking': ie using the intuition gained from experience to build up a machine from piecemeal components, with little overall *structure* to the design. The successful Motorola 6800 memory chip was designed in this way. Its designer spent over a year working on it, and is reputed to have vowed afterwards 'That's the last of the big mothers!'

That anecdote was told by Carver Mead, to whom the designer of the chip went for a course on structured design. This design method involved ideas derived from software design and the structured programming movement. The basis of the structured philosophy is to control complexity by dividing a complex structure into smaller, more manageable substructures. These substructures are determined by looking at the complex object with specific engineering concerns in mind, from which simplified plans for the design can be made. Then the resulting parts of the design can be worked on in parallel by several people, thus speeding up the process of design considerably. Chapter 5 is concerned with what it means to split up a design in this way, and investigates the nature of the different representations used to simplify the design at different stages in the design process.

Some concrete ideas about what makes a good CAD system can be got from knowledge of the structure of the design task (Chapter 3), insight into the levels of abstraction used by designers in different circumstances (Chapter 5) and from the criteria suggested by designers themselves (Chapter 4). The first question to be asked is: Should improvements be made to existing systems, or should an entirely new approach be taken? Systems which are being developed by researchers in artificial intelligence (AI) suggest new approaches to the basic structure and constituents of CAD systems. They help to determine the requirement for a system that can have the characteristics that adequately model the design task and, at the same time, have all the features needed for a good human-computer interface.

In Chapter 6 several new developments in CAD are considered in which particular features could contribute to an 'ideal' CAD system. The advantages and disadvantages of rule-based 'expert' systems, the importance of different forms of representation for design, the feasibility of complete design automation, and the role of the CAD system as a passive tool or as an active advisor are all considered. These ideas contribute to Chapter 7, where a design system is described which shows how algorithmic and heuristic representations of knowledge about electronics can be made to work together in a system with which the user can consult and negotiate in a realistically interactive fashion.

CHAPTER 2

What *is* CAD?

Computer-aided engineering is breaking up the old compartments into which manufacturing industry has been traditionally divided. It is moving us along a different, more challenging path. Much routine work can now be taken over by computer aids, leaving the engineer to apply all his professional expertise to a broader view of the task in hand. They will allow him to use his time and skills to make certain that projects are both properly conceived and executed.

Patrick Jenkin, Secretary of State for Industry,
Electronics and Power, January 1983

The Science Policy Research Unit of Sussex University recently produced a report showing that only 150 engineering companies out of a total of 25,000 in the UK have installed CAD systems.

Computer Weekly, 8 December 1982

The problem of complexity in electronics design can only really be solved with the help of computers. There are two roles which the computer plays in making complex problems tractable: as a tireless slave, performing simple repetitive operations to generate or simulate designs, and as an information source and communication link between designers, providing quick access to design data. Some jobs, especially in integrated circuit (IC) design, would be so time consuming to do manually that they are only possible with computer aids. The computer's role as a communications link becomes important when a complex system is designed by a team, rather than by one person. The philosophy of complexity control is 'divide and conquer'. The application of this maxim to the design of large-scale electronic components has resulted in the idea of a design team, made up of members of several professional groups, eg systems analysts, systems designers, electronics designers, layout designers, etc. Each of these professional groups has different concerns and uses a different sort of CAD tool (see Chapter 3).

The use of CAD

Most CAD tools currently in use are concerned with logic and layout design, though there is an increasing demand for very high-level tools for specification of systems. These different tools run on different types of computer hardware. Tools for logic design (logic simulators) and circuit analysis (circuit simulators) are large number-crunching programs which run mathematical models of circuits. Layout design also requires a lot of computing power, but concentrates on manipulation and transformation of graphic symbols.

Despite the increasing need for computer assistance in producing electronics designs, comparatively few CAD systems are being used in Britain today. Before considering some requirements for a better CAD system, it would be useful to look at some mistakes that have been made by electronics companies in the recent past.

Problems arising from the use of CAD

The industrial concerns which use CAD are mostly giants with larger than average resources. These giants stand as typical examples of firms who were early users of CAD, because they were the few who could afford to buy such expensive tools, or who needed to use them because of the size and complexity of their projects. Shortage of skilled manpower was also a problem for which CAD seemed to be a solution:

> When CAD was originally considered for introduction to some of the British Aerospace design offices, the problem to be overcome was how to maintain a high level of design-engineering activity in a decreasing level of qualified design engineers available to achieve the results . . . It was necessary to utilise the time of the talented design engineers to best advantage, and CAD offered the opportunity to use the skills to the maximum. It allowed a machine to undertake the comparatively mundane tasks of illustrating the design, leaving the engineer to continue the creative tasks.
>
> *Birtles 1983*

Some of the problems experienced by early CAD users stemmed from the timing of the initial decision to invest in CAD tools. Some of the tools they were using in 1982 had been marketed in 1976 and perhaps designed three to five years earlier. Others had been bought in 1981 and were cheaper and more appropriate systems, but the old machines had yet to make a return on the initial investment. The old machines had to be kept working, even if they were incompatible with the new ones. Data was often transferred between machines by physically moving magnetic tapes, which required interface programs to make the formats of the data compatible. This was a cumbersome and time-wasting exercise.

> The first installation ran to about half a million pounds, and I don't care who you are, even if you are a multinational, you have to justify that investment . . .
>
> *CAD manager (from Begg 1983a)*

There was a clash of a sort typical in high technology between the need for the latest, most efficient tools and the necessity of amortizing a financial investment. The two requirements were not seen to be in conflict at first: it was only when it was realized what the technology concerned could actually do, and how fast it became outdated, that it became obvious.

> Unhappy experiences with CAD may range from frustration with investment in time and cost necessary to achieve a successful result, to downright failure, culminating in the removal of the CAD system and reversion to manual methods . . . The principal cause of such disappointments is inadequate preparation and planning. It is remarkable how often organisations that use quite sophisticated planning processes to control their product design and manufacturing operations will fail to apply the same discipline to a CAD project. This even occurs with organisations that are extremely familiar with the hardware and software technologies employed in CAD.
>
> *Green 1983*

On the other hand, it can be argued that the lack of choice in early CAD systems, combined with a degree of what has been called computer illiteracy on the part of electronics engineers led the first CAD users into difficult situations.

There has been an overlap effect in CAD technology, rather like that in semiconductor design: complexity has increased, design times have got longer and market pressures have been so strong that products have been out of date before their designs are finished. CAD systems designers seem to have been keen to reach a very wide market with early systems to suit (loosely speaking) a wide range of applications, but these systems failed to meet the special needs of electronics design. An example of this comes from a designer of ICs, who complained to the author that when he first began to use it, his CAD system, developed from a map-drawing application, gave chip areas in nanoacres. In general, there was a feeling that CAD systems for electronics were not designed for the job, and that this was due to ignorance of the hardware designer's needs on the part of the software designer.

This is a classic problem in man-machine or, as it will be known in this book, human-computer interface design. The human-computer interface is one of the most sensitive areas in software design, and much progress has been made lately by psychologists, ergonomists and programmers:

> Unfortunately, interaction has been until recently one of the neglected parts of many computer systems. System builders have traditionally focussed more on making the system work than on the fact that people need to use it as a tool for problem solving. Moreover, system builders do not easily understand human-computer interaction because it involves so many factors that cannot be pinned down with algorithms.
>
> *Bo 1982*

The problem is deeper than the surface language difficulties which people have with computers. The process of automating a cognitive process often reveals our ignorance of those processes. The designer of an automated system implements the model he has of the task to be carried out. That model may not be as complete or correct as future users (who may be expert in the automated task themselves) need or anticipate:

> All too often, designers and installers of automation systems do not realise the unformalisable subtleties of the work being automated, and therefore do not anticipate the differences between what the systems are going to do and what the people did whom they are replacing.
>
> *Fikes 1982*

For further discussion of the formalizability of problems and solutions see Chapter 5.

Another contributory factor to dissatisfaction with CAD systems is that staff involved in *choosing* a system (usually middle management) are often not familiar with the needs of their staff as designers. CAD seems to some managers to solve an organizational problem without proper analysis of the task involved: CAD is thought to bring increased productivity using less human resources. This is definitely not the case. Only the most simple and boring of tasks were performed by early CAD tools, and even today's work-stations are not capable of replacing skilled designers. In fact, what CAD systems do is to remove drudgery and frustration and to improve product quality by allowing the designer time to be more precise and thorough. Moreover, it improves communication between designers working on the same project and makes modification (and indeed wholesale changes of mind) possible.

For bemused managers, Constantinou *et al.* (1982) gives advice on various types of CAD systems and how to sort out conflicting objectives, as well as a procedure for evaluating the development of CAD systems. Having discussed some background problems, the ground is prepared for making a *user model.* This is the first step in designing a human-computer interface: to discover what it is that needs to be automated, how much of that can be automated, and how the potential user of a system habitually performs the task.

The use of CAD in electronics design
Figure 2.1 illustrates the kinds of design tool used in different electronics design applications. The kind of CAD tool that is used varies between analog and digital applications and between PCB and chip technologies.

digital design		analog
PCBs	ICs	design
logic simulation		circuit analysis
automatic layout routines		
layout verification		
design rule checking		
artwork	mask making	artwork

Figure 2.1 *Areas of electronics design where CAD is used*

Analog applications are declining, but not disappearing entirely:

> The vast majority of semicustom ICs at present are purely digital
> . . . Analogue functions are, however, often requested by custom-
> ers, and there is a growing demand for semicustom analogue ICs.
>
> *Broster & South 1983*

Analog circuit analysis is similar in character to digital logic simulation. In both, mathematical models of circuits are used as a way to test designs before actual manufacture. This kind of calculation is very time consuming to carry out manually, but can be worthwhile when used as a debugging aid. Without it, one has to rely on intuition. This can work in small-scale designs, but does not with large, complex ones.

Analog circuit designers have a hands-on relationship with what they design but this is impossible to carry through in IC technology, simply because the finished product is too small to take an oscilloscope probe! However, the differences in training and attitude to mathematical models between analog and digital designers leads them to a much more lively appreciation of what is going on inside an analysis package than a digital designer has of a simulator. There is a similar contrast between the attitudes of PCB and IC designers: the IC designer uses CAD to solve design problems which are *known* to be intractable without computer aids. The PCB designer, on the other hand, has a much better model of his task. Much more is known about good PCB design, and so the PCB designer can see just how inadequately his program performs in comparison with a person. The IC designer physically cannot begin to develop manual expertise in designing chips.

Conclusions

The tasks that have been automated in CAD systems up to now have been those which computers have been considered to do well: large, complex calculations, simple symbol manipulation and pattern matching:

1) problems that combine the manipulation of large amounts of data with the calculation of complex functions (eg simulation and circuit analysis);
2) problems of conversion from one form of representation

to another (eg logic diagrams to layout, layout to masks);

3) problems that involve boring check-lists (eg design rule checking, layout verification).

These tasks have varying degrees of significance for each technology and in each application. How CAD tools are used in the design process are described in Chapter 3.

How is design carried out?

The design process

Figure 3.1 represents the processing of a design through the design team of a large firm. Note that each of the subgroups within the design team may have had a different education and training; they will certainly have a different style of representing the design. This has a considerable impact on the requirements they each have of a CAD system. A smaller concern would not break down a design to the same degree because the size and complexity of their projects will tend to be less.

people involved	*product*
customer	
systems analyst and designer ⟶	system specifications
project manager ⟶	submodule specifications
electronics designer ⟶	submodule circuit/logic diagram
draughtsman ⟶	layout/artwork
manufacturer ⟶	product

Figure 3.1 *An overall view of the process of design*

The issue of design representation will be discussed fully in Chapter 5. For now, it is enough to say that each subset of the design team uses a different method of representation for a design. Each time the processing of a design progresses towards the manufacturing phase (downwards in Figure 3.1), it is broken down into smaller chunks which can be implemented in greater detail. The number of the steps shown in Figure 3.1 is representative of those observed, but is arbitrary with respect to anything but traditional division of labour within large organizations. In this chapter the three lower levels of Figure 3.1 will be considered, which is where CAD is most often used. At these levels, the size and intricacy of the design become problematic because of the increasing level of implementation detail that is being expressed. CAD is used to ensure that the integrity of the design as a whole is preserved, despite the mass of detail that is being produced.

At the top end of Figure 3.1 is an area in which there is great potential for CAD applications of a more 'intelligent' sort, to tackle problems posed by the specification task itself. At present, such high-level design aids are simply not available. Although it is theoretically advisable to state the behavioural requirements of a machine very carefully before a design is started, the task of determining *all* the relevant requirements is not easy, and engineers tend to rely on a combination of trust in practical experience and a wide knowledge of established technology.

Specifications seldom arrive at the engineer's desk in a very precise form. Here are some extracts from an empirical study of design which show just how informal a specification can be:

VB: 'When you get a new job coming in, what do people ask you to do? Do you get a spec?'

BW: 'Not really. Normally I'm approached by my supervisor. He's approached, probably by marketing, for some reason, perhaps they want a 9MHz repeater designed and developed . . . if we've already got a design, we'd check that it's all in order, but if we haven't got one, we'd go through the chore of designing it . . .'

VB: '. . . do you actually get a bit of paper describing it?'

BW: 'No, I've never had . . .'

VB: 'It all comes by word of mouth?'

BW: 'I've never had a piece of paper saying develop this that and the other.'

'. . . various systems are looked at by the systems people . . . then the system is broken down into submodules as it were and specs are drawn up for these and then that's where I might come in (as a circuit analyst) where someone will give you a spec and say "I want a circuit that will do this".'

'Someone will give me the usual military specification of environment, it will also have a specification of performance . . . We primarily do logic circuitry, so it will have speed, performance, delay through the circuit and its functional

implementation elements . . .'
'The system specification would be a functional equation. It's at that high level, which will specify certain requirements like environmental conditions, the overall requirements for the system, but in the main it is a high level specification that defines pure function.'

'There is some sort of specification from whoever wants the chip, normally a systems designer, who's got to integrate that chip with lots of components to make a product that does something. At the moment those specifications are not particularly good, in the sense that they're not formal specifications, and there's a problem of communication between the specifier and the person who has the task of actually designing it. That's the state of the art, I'm afraid, until the mathematical techniques for formal specification methods take over there, there's always going to be a bit of exposure across that interface.'

Begg 1983a

The specification for some applications in electronics design may be entirely verbal and very loose. It may be concerned with functional description alone, or may include parameters related to the environment in which it is to function. It may be as brief as a few sentences, or be recorded in a roomful of documents. (The specification for British Telecom's System X runs to 150 volumes.) Very seldom, the specification is a real functional equation or a set of consistent propositions in a formal specification language (FSL). A formal specification can be difficult to formulate and to interpret back to the customer. However, the aptness of FSLs to be translated directly into other formal languages which describe lower levels of design confers great advantages, both in terms of computability and verifiability.

Even though the specification may not be very precise, the engineer's notions of what sort of a machine he is able to design are. He has a definite idea of a 'practical machine', which stems from his working experience and involves the use of *ready-made parts* that are to hand.

Ready-made parts vary according to the level of description considered and the technology involved. A customer may require a fairly minor modification to an existing machine, in which case design activity is minimal (eg tightening up environmental parameters for a special purpose, as in the machine described by the systems designer earlier). An analog designer may decide that large units (of the scale of the unit amplifier) from old designs can be included in a new project, just as a board designer can include chips or smaller components (eg capacitors and resistors). A chip designer has a more difficult task, for each chip design is, in a sense, a new beginning. The equivalent for the chip designer of using already existing ready-made parts is to use *cells* or partial designs. These are small-scale functional units (which may be repeatable to form large areas, eg of memory cells) but they are not so well-defined functionally as those components described above — they are at a lower implementation level. The arbitrary joining together of cells at this level might have unpredictable effects in the behaviour of the resulting machine. This means that the LSI layout designer does a lot more design than his counterpart laying out boards: the pieces of the jigsaw assembled are much smaller and more enigmatic.

To make the design of LSI chips more tractable, several strategies have been formulated. A full custom design, in which a design is created from scratch, can take upwards of six months in the design stage, which is too long for any but the largest volume production runs to make it profitable. Semi-custom design involves the development of standard cells. These are small sections which can be fitted together to make larger units, a kind of prefabrication of partial designs.

Controlling complexity and saving money are closely related. The power of prefabrication of parts has been proven in other fields, such as mechanical engineering and architectural design, and there too, the technique enables the design of complex structures at an affordable price.

Another approach to LSI design is to create a modular array which can be adapted by connecting the modules in different ways, taking advantage of the fact that the last process in manufacture is the application of a metallization mask. Any modifications to the design are made at this point. This gives flexibility to a standard design. There are variations on this technique, the two most commonly used being the *uncommitted logic array* (ULA, or gate array, or masterslice), in which components such as flip-flops and multiplexors are created by connecting together groups of transistors already extant in the ULA, and the *programmable logic array* (PLA), in which smaller areas of chip are established which potentially form any combinatorial function on a set of input lines: they are programmable in that one can choose which output lines to read, and thus which function is computed. Both of these approaches are ways of structuring design for large-scale devices with many modules on each chip.

Problems in chip design are very like those met in the design of software. Essentially there are blocks of program or chip area which perform specific functions and which have to be coordinated, controlled and supplied with the right data at the right time. Mead & Conway (1979) have promoted the idea of *structured design* for VLSI hardware. This approach has been very influential, since it provides both a new and more appropriate form of representation for VLSI design and a set of basic rules which are meant to reduce the incidence of design errors.

The rest of this chapter will be concerned with the CAD tools associated with circuit, logic and layout design, and the human tasks which accompany their use in each step of the design process.

Circuit analysis

This covers the many mathematical techniques used in the design of (especially) analog circuits. Computer programs using these techniques were some of the first CAD tools. Circuit analysis programs are circuit simulators into which a wide variety of ideal models of components can be plugged. As was mentioned earlier, analog design is a declining field, and not much use is made of circuit analysis programs for this application. ICs are so complex that it is difficult to model them in great detail. However, programs like SPICE are widely used and constantly being improved.

Logic design

Logic design has the paradoxical reputation of being a black art, while at the same time being the most heavily computer-aided high-level design concern. The logic designer possesses the most general expertise of the design team. He has to be aware of factors which influence the design at the logic level from both above (at the system level) and below (at the layout level). The job of the logic designer is difficult to describe: it is necessary to take many different sorts of parameters into account. Once one has a specification of some kind, the process of logic design is one of translating the specification into logic functions, whilst bearing in mind the sort of implementation technology which is available.

Logic simulation

The main CAD tool used by logic designers is the logic simulator. This is a complex suite of software with a purpose of circuit verification and functional checking. The logic design has to be translated from a graphical representation (in most cases, with some recent exceptions) into some design language, a textual representation. For the purposes of verification, logic circuits are sometimes described in terms of their connectivity alone, so that basic errors such as open circuits can be detected. This kind of description is sometimes called a *netlist*. Electronics engineers are not fond of most design languages, since they impose lengthy and tedious translation tasks at a time when the engineer feels that he can least well spare the time. However, design languages are necessary because they form the basic representation of the design for the CAD system itself. This basic design information may be compiled and recompiled into different forms, but it is (potentially) a reference point for all design tools.

Functional checking is a far more interesting problem than plain verification. It means ensuring the logic functions implemented produce the appropriate input/output (I/O) behaviour, which is monitored at predefined test points. The format of the printed report is under the engineer's control in some cases, so that sets of signals can be grouped together and labelled in a meaningful way. It is not possible to simulate all the possible conditions of the circuit for any practical design. This (common) problem is overcome by the engineer designing for testability. In order to design in this way, it is necessary to have an overview of the intended function of the machine and its intended implementation. The design engineer's contribution is crucial to the success of the final design, since he produces a vital link in the chain of transformations constituting the design process.

LOGIC TO LAYOUT

The logic designer is not the sole contributor to the process of design. One imagines, as a naive observer of engineering design, that the drawing office would be a place where people generally copied, or at most, expanded, the designs produced by engineers. In fact, the production of a layout, even at the comparatively unsophisticated board level, is a translation every bit as complex in its demand for different sorts of knowledge as that carried out by the electronics engineer. The form in which the design reaches the layout designer may be every bit as informal as the specification that comes to the engineer:

> VB: 'What form is that circuit design in before you get it?'
> Layout designer: 'Before we get it it's usually on the back of an envelope.'
>
> *Begg 1983a*

This exchange sounds like a joke, but it isn't. The engineer considers his rough sketch to be sufficient to suggest a layout. He does not consider it to be his job to produce neat drawings.

When the engineer is required to contribute to the design team in a more precise and well-documented way, he often complains about it. The author recently interviewed a representative of the CAD user's group in a company. The engineers there are required to render their designs into a design language and input them directly to the CAD system. This makes the draughtsman's job a lot easier and cuts down the number of errors in basics, such as connectivity. However, the use of a design language as input is resented by engineers here as elsewhere:

> The design language (DL) is mandatory, you have to use it in designing a board . . . that means that right at the busiest time of the design cycle, he has to set to and produce a DL file and he has got to be 100% right and 100% complete . . . The connectivity part of DL that says what's joined to what is dead easy. Once you know a few simple rules you can often give it to clerical staff, or we would do if we had them . . . and we are talking about 40, 50 or 60 pages of coding sheets in some cases . . . An engineer who is working flat out, overtime and weekends sometimes, to get his board working as a design has got to code the whole thing up as well . . . That means that just the sheer fag of writing out those sheets is umpteen man-hours . . .

Another problem arises from the fact that the engineer in this case is a casual user:

> Any one engineer is going to be doing one or two boards in the course of a year at the most, so having to learn DL from scratch (each time) . . . is wasted knowledge.
>
> *Begg 1983a*

There are advantages in engineers using a design language, for other members of the design team: since the process of translating the circuit into computer-readable form creates design data which can be used immediately by the layout designer and his CAD tools. Once the circuit is in the computer it can be kept in a verified form: if it has to be repeatedly re-coded, human error is bound to corrupt it.

In most cases the design language is a description of the connectivity of the circuit, in effect a list describing a network of logic gates. A particularly boring job for an engineer, but a very necessary one as things stand, is the packaging and labelling of gates. The layout designer and the logic designer have to confer in order to ensure that the design is progressing along the right lines. If gates are not labelled, it is difficult to keep track of exactly how the logic has been implemented as a layout (either with CAD tools or manually). Logic designers may be keen to see that groups of gates have been packaged and placed in a particular way. Another reason for labelling gates is so that errors can be isolated and traced. The job of labelling is boring and time consuming, but unless the whole translation process is carried out using a computer, the task falls to the engineer.

Manual layout procedures

Manual layout design is one field in which CAD has been used to solve most of the problems inherent in the design process. Manual layout is time consuming and demands great manual dexterity and patience: each layer of a PCB, for example, would be represented on a separate layer of clear film, and the components and their interconnections represented by self-adhesive symbols and black crepe tape. This kind of documentation was perishable (the tape started to creep after a couple of weeks) and easy to lose. For integrated circuits under 10,000 gates manual layouts covered whole walls and were virtually impossible to verify. CAD speeds up the layout process by cutting down the time it takes to correct errors. In addition, it provides a stable, verifiable representation of the design which can be accessed by both people and programs. In the following sections the contribution of automatic layout routines to the finished design is examined.

Automatic layout routines – PCBs

PRELIMINARIES

The author recently interviewed several PCB designers, the automatic layout routines used by the interviewees were of two types, placement and auto-routing. Before these can be used, the designer has to ask several questions about basic parameters. Here is a description of the first steps in draughting a PCB:

> The first thing you have to find out is how big a PCB you're going to have to use . . . The second one, you try to assess whether it's going to be a single layer PCB, double layer, or multi-layer, if it's multi-layer, how many layers it's going to be. You assess how big the components are, ie will the PCB be big enough anyway, or will you need two . . . Then you start looking for specific constraints, like, are there two components that mustn't be close to each other, is the power supply particularly touchy, have you got to keep that separated, is there any screening of components. Generally go through the circuit diagram, in the case of a PCB, looking at it and talking to the circuit designer, seeing what his feelings are, getting an assessment of the overall picture of the thing.
>
> *Begg 1983a*

Consultation with the circuit or logic designer at this first stage is very important to the success of the design, especially when automatic verification is not available. The layout designer makes the final decision as to which ready-made components go on to the board.

The catalogues and data books used by designers are invaluable. They are often lost, stolen or misplaced, sometimes dog-eared and coffee-stained, but it is essential that they contain the most up-to-date information about the components that exist and which are available for use. Some CAD systems have on-line component libraries, but it is difficult to keep them up-to-date, since new parts are added and old parts drop out of use before the catalogue lists can be entered into the database. Most designers agree that an up-to-date on-line component library would be a greatly appreciated facility in any CAD system.

PACKAGING

Packaging has been mentioned before in the context of logic design. Packaging, as was noted earlier, is usually the layout designer's responsibility. Because the engineer is mainly concerned with logical function and not physical structure, individual gates on the logic diagram have to be reorganized into rational shapes. This can be done using a computer, but the computer cannot be relied on to always package gates in a way that seems rational to the designer. The programs are generally as sophisticated as they can be, but when it comes to packaging they do not use the same information as the draughtsman. The program can only operate on the netlist. The draughtsman can see the whole board and use his often superb pattern recognition abilities to detect features that are not representable in the netlist form. Besides, criteria concerning the function of gate packages are used to make packaging decisions. The designer uses his experience and intuition to model what the completed design will look like in order to package gates appropriately.

AUTOMATIC PLACEMENT ROUTINES

Input to an auto-placement routine consists of a list of components and their connectivity, either drawn directly from the engineer's rough sketch, or pre-processed by the CAD system. The placement program then divides the board area into a matrix of slots into which only one component can fit, and places them one by one, starting with the most connected component, which it places roughly in the middle of the board. More and more sophisticated placement algorithms are continually being suggested, but those available at the moment are seldom used, since connectivity is not the only factor bearing on placement. 'Bad' placement by a program is more time consuming in terms of correction than the slightly slower initial pace of 'correct' placement by a designer. Correct placement can only be judged with reference to the sum of a designer's knowledge and experience.

AUTOMATIC ROUTING

Once the placing of the components has been decided and connectivity relations are established, automatic routing would seem to be a fairly simple task. However, there is a further set of constraints, the design rules, which first must be taken into consideration. These are a codification of the limitations of the technology, a summary of what the designer must not depict because the manufacturing process cannot produce it. (For further discussion of design rules and design rule checkers, see page 30.)

There could be between 15 and 150 design rules laid down for a project, those for routing would cover width of track, clearances, etc. These have to be incorporated into an already complex routing algorithm. It also has to be designed to make the most efficient and economical use of metal. Users applaud the quickness of autorouting as it is compared to the laborious task of manual routing, but it is not a completely automatic process. Some PCB autorouting packages are interactive to a large degree, and allow the user to set up the board in such a way as to leave fewer problems for the autorouter, and allows him to finish off when the autorouter can complete no more tracks. Most autorouters will complete up to 75-90% of a board, but the remaining 10% of a routing task will often be the most difficult, because the automatic routine leaves little space for the last wires to be run.

Using an autorouter isn't just a matter of pressing a button and waiting. The designer has to set up the data for the program, then finish off a nearly complete design. The machine complements the designer's efforts rather than replaces them. This interactivity brings advantages over manual layout besides mere speed of execution. (Though it pays to note the fact that it takes a plotter only ten minutes to produce a design that would take a tracer three and a half days to draw; however, it can take a month to get the relevant design data into the computer.)

Not only can errors be detected and corrected quickly, but the flexibility of the graphics terminal display allows one to think about making moves that would have seemed impracticable to the manual draughtsman, such as making global corrections to occurrences of one cell shape, or inverting sections of a design, or turning blocks through 90 degrees.

Automatic layout routines – ICs

Moving from CAD tools for PCB layout to those used for chip design is a major step. ICs can be laid out on similar graphics systems to those that have been discussed, but dedicated, single-user work-stations are preferred for their flexibility and versatility. IC design is becoming more and more like computer programming. The transition from logic diagram, perhaps via a sticks diagram (an intermediate form) to the description of the masks themselves can be automated (eg the INMOS IC design tools). A considerable amount of effort has gone into design automation in this area and much expertise gained from software engineering has been applied. Automatic chip designing programs are called *silicon compilers*.

> A silicon compiler translates a behavioural description of a function into a set of geometric images which can be used to fabricate an integrated circuit that performs that function. A sophisticated silicon compiler has the potential of allowing the design of very large circuits based on the same methodology that allows modern software compilers to generate very large complex programs.
>
> *Rupp 1981*

Silicon compilers are specialized programs which deal only with a limited range of types of design (eg signal processing) and are not general solutions. They are useful when a design has to be produced quickly and when there is little concern for minimization of area.

Design rule checkers

Design rule checkers (DRCs) are universally considered to be the most boring kind of CAD tool available. They are basically a codification, in the form of a set of rules of the limitations of the manufacturing process being used to create the chip. A design rule set is prefaced with a set of definitions for the variables mentioned. The rules look something like:

```
RULE MIN WIN            [rule name]
    FAIL ' ' IF WIDTH (CUT) < 6   [rule body]
END
```

which returns an error message of some kind, perhaps textual, perhaps a graphical pointer, if a feature is smaller than some predetermined size, and hence unmanufacturable. This is obviously an extremely primitive and simple rule. Most other rules are more complex Boolean statements, but this representation does not give an accurate enough model of the technology being used. DRCs produce a large number of spurious errors: errors are signalled when in fact the configuration shown is perfectly 'legal'. In some cases it is possible to overcome the problem of spurious errors by redesigning drawing standards, but this is an unsatisfactory measure. In a good CAD system, the tail should not be seen to wag the dog!

Many DRCs are, in addition, painfully slow: so slow that one often doesn't know whether or not the program has hung up. If more than a few errors are discovered in a design, the graphical pointers displayed in order to locate errors become very difficult to understand. They are often good examples of bad software design: the programmer has spent all his energy on optimizing the internal processes of the large search program, and neglected the human-computer interface.

Mask making

The design is as complete as it can be by the time it has passed through the DRC. If the chip is fairly small and if it is very important to get it right first time, a circuit analysis program such as SPICE may be applied, but further simulation is usually deemed a waste of time, and physical testing is the final step. But before then, another CAD tool is used to generate *mask tapes* from the symbolic layout which the designer has produced. These are digital descriptions of the shapes needed for the different masks for each deposition, etching or ion implantation level on the chip.

These final tapes are in a form appropriate to the implementation technology, eg data to drive a plotter generating artwork for photographic reduction, or an electron beam which directly etches a pattern on to the wafer. The link between CAD and computer-aided manufacture has not been exploited to its limits as yet. Information flow between the design team and the mask and chip manufacturers is one-way. It would be useful to establish some form of feedback: chip manufacturers are setting up measurement and testing systems with large databases which help to monitor and control manufacturing processes. Design faults can be detected at an early stage, before packaging and final functional testing of chips. If these faults are either so bad as to require complete redesign, or so minor as to be cured by alterations to one or two masks, then the earlier that the information is directed back to the design team, the better. However, it is a considerably better idea to design for functional testability and in a spirit of preventing design faults rather than arranging for them to be 'cured', since cure at this late stage is likely to be prohibitively expensive.

Conclusions

From this description of the design process it is possible to see how important CAD is to electronics design and in which parts of the design process CAD tools are used. In suggesting improvements to CAD systems it seems that the two equations 'Time is money' and 'Skill is money' have to be carefully balanced. Should one opt for methods of complexity control which affect early stages in the design process, such as producing tighter formal specifications, which allow for fewer design decisions? Or should one provide tools for the later stages of design, which remove the drudgery from logic and layout design, but allow the designer to deploy the skills accumulated during his education and professional training?

In designing an integrated CAD system, ie one that can accept design data from and be used by all the members of a design team, it is useful to know which elements of the problem are intractable before going on to further analysis — or to suggest solutions. It appears, from Chapters 2 and 3, that there are three types of problem encountered in designing with CAD tools. First, there are those created by bad software design. These problems are easy to remedy: basic principles for creating a user-friendly human-computer interface have been defined. Second, there are problems which are impossible to remedy by simply redesigning a few computer programs, eg the effects of the kind of management practices mentioned in Chapter 2, which put emphasis on immediate cost, rather than on long-term planning, which are symptomatic of the unrealistic expectations that some people have of computers. Third, there are the real, hard problems posed by design itself. These will be introduced in Chapter 4, along with a summary of the requirements designers in this field make of a CAD system.

CHAPTER 4

What makes a good CAD system?

A CAD system should live up to its name: it should help the
designer in every way possible, not put obstacles in his way.
Since it is necessary to use computer aids in settings where
projects are large and complex, it is safe to assume that more
than one designer will be employed on any one project.
Therefore the CAD system has not only to provide help in
performing specific tasks, but must also provide a good
communications link between designers doing different sorts
of tasks. The human-computer interface in this case must be
reconstrued as a many-to-many relationship, rather than as
one person dealing with one machine.

Communication through the human-computer interface
There is evidence from the way that existing CAD systems
fail that attention to this communications aspect of the
system will save much time and effort. Work is often duplica-
ted in more than one stage of the design. For example,
calculations done by computer in one department may be
needed to be done again manually in another if the format of
the data from the previous step or the medium on which it
was stored are not compatible with the CAD equipment
available. Duplication also appears when graphical data has to
be put into computer-readable form (eg digitizing circuit
diagrams). This appears to the design team to be unnecessarily
repetitive and must inevitably introduce human transcription
errors. Progress has been made in Japan and elsewhere in the
computer interpretation of diagrams and sketches (Latombe
1978, Mackworth 1983) and continues to be an interesting
and important area of research. Overall, it can be concluded
that in a good CAD system the output of one design tool
should be compatible with the input of the next in line (eg

automatic layout routines should always and automatically generate input to the available simulators, verification algorithms and mask making programs).

It is well known that bad human-computer interface design can cause intense frustration for users. However, it seems the lessons of ten years of research into dialogue design for interactive computer use have not always been learned by those who design CAD systems. In the author's experience the most frequent complaints made by CAD users have been about menu systems, on-line documentation and error messages, where the application of a few simple principles of software design were all that was needed to render the system far more user-friendly. It may be difficult to strike a balance between output that is too complex and that which is so simple that it makes the user feel patronized. Incomprehensible and sometimes downright rude protocols have created an atmosphere of distrust and even fear for some communities of CAD users.

Often the lack of support and inadequate user modelling in CAD systems lead people to doubt whether they have any control over the machine at all. This leads to difficulties in learning (if the idea of using the system at all remains attractive) which are exacerbated by the use of different kinds of computers, operating systems and design tools in different departments of the same organization. If it is worth investing in a large CAD system, it is worth ensuring that the system is sufficiently well integrated to be usable by designers, who will not be computer experts and who are not interested in learning a new profession. The concept of an *integrated system* sums up most of the criteria for good CAD systems that have been suggested to the author. The integrated system is one in which information about the design flows freely between all the members of the design team and can be employed where necessary using the full range of CAD tools.

In this regard it is interesting to note that a human-computer interface which is so complex as to require special expertise to negotiate creates a powerful elite within an organization. Information can become inaccessible to most people once it has been 'put on the system'. The author found an example of this in a company that had trained a limited number of designers to operate their CAD system. These people had total control over the production of designs via the flow of design information through a bottleneck in the drawing office. This control gave them the power to hold the company to ransom which, in a time of industrial strife, they did. The company subsequently invested in a more user-friendly CAD system and more training courses for its employees, a measure which helped to overcome considerable resistance to new technology among the remainder of their staff.

Bearing this in mind, the reader will endorse the prescriptions of research in this area that interfaces of this kind should be supportive (crash-proof) and easy to learn. More progressive interface programs can tailor themselves to the needs of individual users (eg by providing a dual mode menu system that allows the user to suppress 'teaching texts'). This not only increases the user's sense of control over the system, but is seen as providing the necessary quick return on a company's investment by getting the system used as quickly as possible.

Designers who are dissatisfied with CAD tools can express this by underusing the tools at their disposal. In some companies, for example, logic simulators are considered to be too time consuming to set up: the faults that the simulator would otherwise detect are quick to patch. Because the human-computer interface is poor, the use of a computer aid which guarantees good results is considered to be more trouble than it is worth. If CAD tools are to be used in industrial settings, they must provide quick results, as well as good results, otherwise they will be neglected in favour of 'dirty', but time saving, engineering practices. (However, the heuristics which allow designers to design for modifiability will be of great interest in automating design, see Chapter 7.)

A good CAD system is one which provides good, relevant information, fast. A counter example is of a system whose database was not updated regularly enough and whose users had to regularly make trips to a library in a remote part of an establishment in order to consult data books. The data books were not allowed out of the library because 'they were always disappearing'. An out-of-date component library is useless to a designer.

The CAD users' needs
In Chapter 3 it was noted that computer aids for the higher

levels of design are sadly lacking. In a good CAD system there should be facilities for using higher level design languages (especially for VLSI design) so that the designer does not get so bogged-down in detail. In order to use the communications links provided by an integrated CAD system, the design ought to be put into the system at the earliest stage — from the point of view of maintaining good documentation, if nothing else. However, this raises the problem of how to manage access to the design at different stages. Not all members of the design team require access to all levels, and it is important that some kinds of design information are secure. In general, read-only access should be given to CAD programs and files for both processing and query or instructional purposes, but an individual's current work and objects such as standard cell libraries should be strenuously protected against tampering or accidental loss.

However carefully a CAD system designer tries to echo CAD users' needs, there are difficulties which arise from taking the user's part too enthusiastically. Current users' ideas of what a good CAD system will look like are conditional on what they predict the capacity of computers in the future to be, based on their own experience and on currently held beliefs and apprehensions. One of these apprehensions is that use of CAD tools will replace many members of the design team. This fear of encroaching automation may be temporary in its effects, wearing off as users become accustomed to using the computer as a tool.

The threat of job loss due to automation has not affected this particular field. No one has yet succeeded in producing a commercially successful automated designer, and even the successful draughting tools have by no means replaced skilled people. The design methodology of electronics is itself undergoing constant development, and this hardly keeps up with the frantic pace of the semiconductor industry that it serves. The task of designing a good human-computer interface for a CAD system is complicated by this fact. There is a constant problem of trying to produce familiar looking tools to do a job which is new enough (at least as far as LSI circuits are concerned) to have no traditional design methodology.

In addition, one is faced with a problem analogous to that described in the design of semiconductors, where design times are overhauled by new developments in technology. It was possible to buy CAD systems which met the users' needs at the time the study was undertaken, but systems had to be bought at a time when CAD was less well developed. In designing a better human-computer interface one has to reconcile the user's fear of the new with the demands of a brand new technology, while keeping in mind the rate at which the technology in which one is designing will overhaul presently available systems.

There are real dilemmas posed by the rich ferment of ideas in electronics design: for example, many electronics designers would like to see CAD tools which provide more and more precise simulations of logical and electrical behaviour, as if the real hardware were there to test with an oscilloscope. However, other designers are convinced that it is only possible to make progress in complex designs by using highly idealized simple models of devices. To which set of design experts should the CAD designer listen?

The latter is an attractive solution for the CAD system designer, since most simulators are considered to be very inaccurate. The problem of simulation is horrendously complicated, containing some sub-problems which are mathematically intractable, and others which are theoretically feasible but become uncomputable with large numbers of variables (Lewin 1977). Thus, users who ask for more realism in simulations are, for technical reasons, unlikely to get it. Besides, in hardware as well as software design, the structured design approach is becoming more and more popular. Designs can be made more tractable and testable by breaking up the design into modular units and by abstraction (see Chapter 5). This approach is by no means in widespread use in all sections of industry. To envisage CAD tools that introduce structured design techniques requires that one designs an interface that educates the user rather than models the skills which had previously been used.

When factors like these, not usually considered in studies of computer users, are taken into consideration, it becomes obvious that one cannot always provide exactly what users say they want from a system, because what they want is contingent on the value system arising from their current social and technical context, which may well change with time. For example, the draughtsman may not actually want a realistic simulation of his drawing-board (which might bring with it disadvantages as well as advantages): he just needs to be able to handle a CAD system with the same degree of ease with which he can manipulate a drawing-board. Likewise, the manager may express a desire for a fully integrated CAD system, but be motivated more by the apparently magical promise of automation than by an informed assessment of the needs of his staff.

THE USER MODEL

A fundamental concept in considering improvements to the human-computer interface in any system is the *user model*. This can be thought of as a list of parameters drawn up by the system designer which indicates how users think about the task in hand. There are several ways to construct the user model: from the point of view of the user (a model of how the user thinks the job is carried out and how he imagines the computer to be of help) and from the point of view of the computer (a model of what the user is likely to do, and what will be demanded of the computer). Making improvements to CAD systems of the future will hinge on the quality of these models.

How should one go about creating a user model for the design team as a whole, with the idea of designing an integrated CAD system? Some members of the design team do not have access to CAD tools as yet. Should their needs be considered? We have seen that CAD attempts several types of task, each tackled by a different member of the design team. Electronics engineers require fast answers to the complex calculation problems of simulation. Draughtsmen need fast access to well-kept libraries and automatic generation and checking of reams of low-level design information. At present, these requirements conflict so badly that completely different arrangements are made for the provision of different sorts of programs. This leads to incompatibility and lack of integration in CAD systems.

THE USER PROFILE

In order to get a good overall view of the qualities of the design team the author has borrowed the format of a *user profile* from Maguire (1982). This approach considers attributes of the intended user of a computer system according to: education, experience, interests, time pressures and manual dexterity.

Members of the design team are mostly educated to technician level, often having served an apprenticeship, not necessarily in electronics engineering. Most of them have a general non-academic engineering background. Very few of them have any experience of computers and computing, though this is likely to change. Highly qualified and skilled people in this area are very rare indeed.

If the designer/draughtsman is older (over 30, say) he is likely to have had work experience in a drawing office or at a technician's bench. If younger, he may have no previous experience of the design and making of machines outside college work — in other words, awareness of the product of the design comes only through the CAD system.

The team's interests are sharply divided between those of finishing a job quickly and cheaply and producing a well-engineered product which looks good (at every level of the design). Balancing these two factors is called *engineering trade-off*: a successful designer is one who can achieve a good balance. To achieve this effectively requires the maximum awareness of the nature and purpose of the design. The information to fuel that awareness comes through the CAD system.

Perceived time pressures depend on the individual's reading of the terms of the engineering trade-off. Non-engineering management push designers to work ever more quickly, designers insist that CAD tools can only speed up production by so much. The CAD system complements the skill of the designer, not replaces it. There is, however, a growing awareness of the need for empirical study of exactly where the bottlenecks are in electronics design.

The people who use CAD systems are in some cases actually selected for manual dexterity, on the basis of their talents as draughtsmen. It was thought that the ability to do technical drawings with a pencil would be a skill related to operating a CAD tool. However, it is difficult to say how the tasks of drawing at a drawing-board and operating a digitizing tablet or using a graphics editor with a keyboard or mouse are related. The physical skills are different, but the cognitive skill of three-dimensional realization underlies them both.

This user profile tells us something of what the interface to a good CAD system should be like:

1) It should be easily understood by hardware engineers, using *their* concepts and *their* jargon.
2) Since all the information available to the designer is likely to come via the CAD system, it would be as well to provide access to as much of it as is possible.
3) If it is accepted that the speed of the actual 'creative' part of the design is unlikely to be increased other than by what will probably be a suboptimal automated designer, then speed of information retrieval, verification and simulation must be increased (see Chapter 6).
4) Since the people who use CAD systems use drawn figures in their work in preference to text, graphical input should seriously be considered at every possible point.

Another way of looking at the design team as users is to consider factors which are likely to determine the kind of system that they will be using. For example, task complexity which is at a high level for every member of the design team. CAD is certainly not an everyday clerical practice. However, interfaces often fail to reflect the way that this complex task is performed. Hierarchically ordered menus can over-determine the sequence of tasks, whereas the free form of under-determined task sequences can lead to confusion (see Figure 4.1).

Another factor likely to affect what demands are made of the system is how frequently the design team uses it. This parameter varies between members of the team. A systems designer might never see a CAD system. An electronics engineer uses a simulator perhaps once in three months. A draughtsman can spend up to seven hours a day in front of a visual display unit. However, it is worth bearing in mind that the casual user may be by no means naif and the regular user (in the case of the author's surveys) quite unsophisticated so far as computers go (cf. Maguire 1982).

Adaptability to computer processing methods also varies between members of the design team. It is not obvious why anyone should be asked to adapt themselves to a computer. A car that was designed in such a way that one had to adapt oneself to drive it would probably be illegal, besides being considered a bad design. However, the adaptability of CAD users does vary. Older users tend to learn the skills of CAD tool use very slowly parrot-fashion, having learned by cruel experience that the computer can be merciless to a naif user, losing data, wiping out hours of work, failing without apparent reason, etc. The model of the system's operation held by these users is very sketchy (many frankly admit that they have no idea how the machine works, even after several weeks of training courses) and thus their predictive power with regard to the computer's performance is very low.

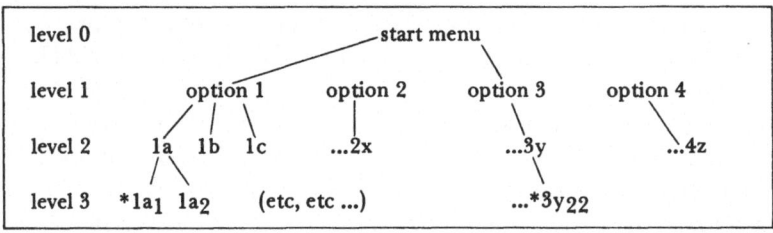

Figure 4.1. *Over-determined, hierarchical menu structure: progressing from option 1a₁ to (say) option 3y₂₂ in such a system is a slow procedure involving traversal of the whole height of the tree; unless the menu is exceptionally well thought out, it is probable that one has to make this difficult move quite frequently*

Those who are new to the job of CAD operation, and the young, tended (on a purely subjective assessment) to treat the highly coloured and mobile graphics at the terminal rather like a game of *Space Invaders*. The responsiveness of the system motivated them to 'play' more with the machine and, through this increased interaction, to learn more about its operation (cf. Carroll 1982). Some qualities of the graphics output of current CAD systems can thus be judged favourably.

However, the pre-video games generation might have benefited from a display which behaved more like a real drawing-board, in that what was drawn stayed there until it was erased, and the logic designers from simulation output whose columns were grouped meaningfully, the least. As was noted earlier, there is a great need for graphical I/O in CAD. A typed list of connections is easily misread, but a drawing of an open circuit can be instantly diagnosed by the human designer's excellent pattern recognition systems.

Suitably structuring a data display increases the speed of interpretation and reduces the error rate.

Stewart (1979) in Maguire (1982)

41

This benefit is the result of using the user's model of the task in hand in the display of data. He sees what he expects to see and he has control over what he sees. The complexity of the task to be performed needs to be ameliorated in every possible way, but this is not to say that one should mimic every aspect of the non-aided task in the CAD system. For example, a draughtsman may not respond well to a drawing-board simulated in every detail, because the simulation may carry with it some of the drawbacks of the manual task: what he expects is the same degree of ease with which he can manipulate a drawing-board, plus the added extras of CAD.

The concern of most psychologists working on issues related to the human-computer interface has been with ergonomics and dialogue design. This sort of study involves focusing-in on a limited area of the human-computer interface. But the human-computer interface has many dimensions and there is more than one way in which it can be analysed. Let us consider one of its dimensions to be 'human senses to program function' (see Figure 4.2).

human senses	computer peripherals	pre- and post-processors	compiler/ interpreter	program function
'shallow' >>'deep'				

Figure 4.2 *The dimensions of the human-computer interface*

Psychologists have been concerned with aspects of the human-computer interface at the 'shallow' end of the scale. 'Deep' concerns in systems design have been left to computer scientists. This procedure has worked well with simple programs and unsophisticated peripherals. In order to design a better interface one has to consider how the program is presented to the user and how the user interprets the program. What if the program is so complex that its output is virtually uninterpretable? There is no sense in just patching it up with a good but shallow interface design, making sure the environment is comfortable, providing flicker-free screens, ensuring response times are quick or slow enough, arguing about the relative merits of text and graphics, etc. One has to consider why the program's output is uninterpretable: what aspects of the output can be made more cognitively digestible as well as perceivable?

Schemes of communication

Within the field of electronics engineering, CAD presents many nearly insoluble problems to the programmer. There are difficult questions of complexity and formalization over the whole field of simulation, for example, which have not yet been answered (see Chapter 6 for further discussion of this). This chapter has pointed out the complexity of the design task and the necessity for CAD tools to be understandable in order to be helpful. The complexity of the programs and the complexity of the design task call for much (pre- and post-) processing of the information that passes through the human-computer interface.

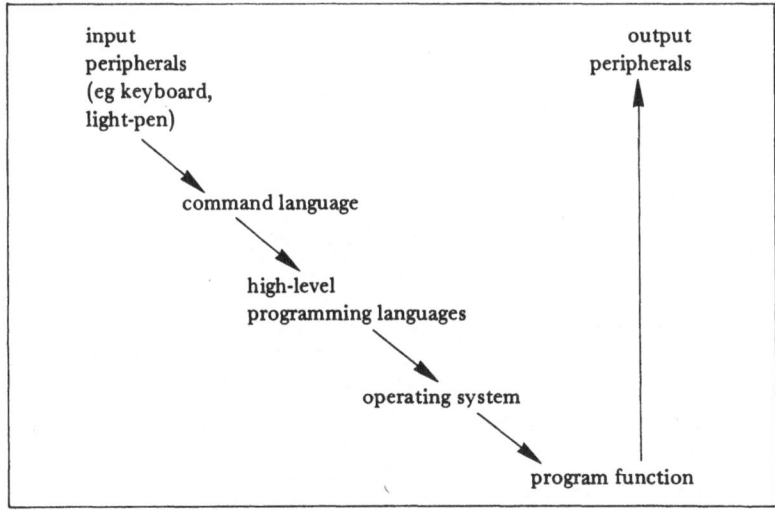

Figure 4.3 *Levels of translation in human-computer communication*

The human-computer interface has been expanded: it is no longer a single channel, nor is it concerned with a single translation process. Figure 4.3 represents a system with a single channel of communication between user and computer, and with a limited number of levels of translation between computer languages. Figure 4.4 shows a system more representative of today's CAD systems, with more peripherals and more layers of different types of language between user and computer.

When this level of complexity is attained in a system, there must be concern that the programs which manage and direct the processing and flow of information are acting in a way which makes using the system easier, not harder.

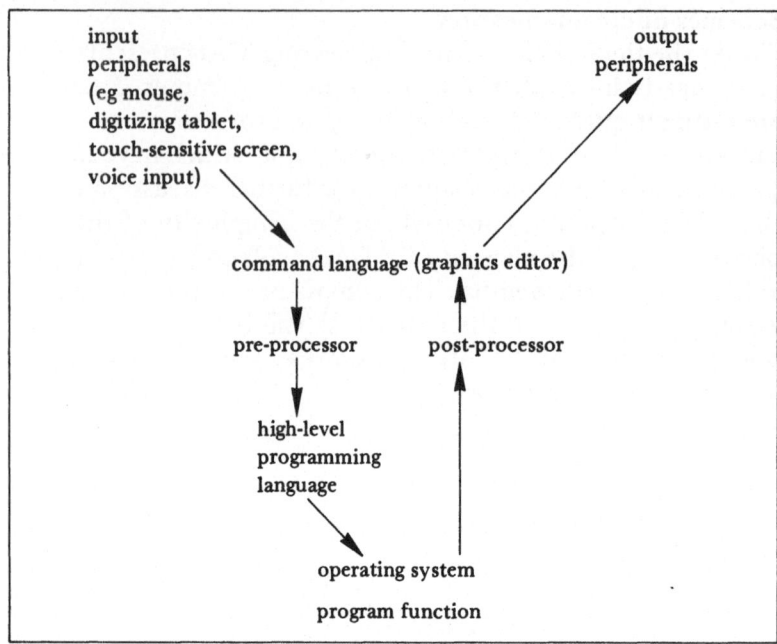

Figure 4.4 *A more complex scheme of communication*

In order to decide on a strategy for organizing and communic- ating with complex programs, the structure of the design task will be analysed further in Chapter 5. This should provide a task model which will help in the design of pre- and post- processors, and suggest to the CAD systems programmer ways to develop different types of CAD system.

Design representation

It can be shown that a mathematical web of some kind can be woven about any universe containing several objects. The fact that our universe lends itself to mathematical treatment is not a fact of any great philosophical significance.

Bertrand Russell

Having determined some of the CAD user's needs, the question arises as to how much of the design process can be automated. At first sight, the prime parameter in such considerations seems to be the nature of the representations used at different stages in design, some being oriented more toward use by human designers, others only being used by computers. However, current movements in electronics design methodology and design automation have highlighted the deficiencies of traditional methods of representing electronics designs, both for people and for computers. This has lead to a vigorous blooming of many varieties of design language for use in CAD. Using as a starting point Herbert Simon's analysis of design as an 'ill-structured problem', a model of design will now be described in the hope that an understanding of the basic objects and operations of the design process will help to distinguish helpful varieties of representation from the underbrush of design languages.

Problem formalization
Over the last ten years, many researchers in AI have come face to face with the task of trying to formalize problems in order to produce intelligent problem-solvers. Design problems have proved especially difficult. Simon (1973) characterizes design (specifically architectural design) as an ill-structured problem, ie one which is difficult to formalize and thus difficult to solve, especially with man-made problem-solvers.

If we look at his criteria for 'well-structured problems' the relevance of this concept to electronics design becomes clear.

1) There is a definite criterion for testing any proposed solution, and a mechanizable process for applying the criterion.
2) There is at least one problem space in which can be represented the initial state, the goal state, and all other states that can be reached or *considered* in the course of attempting a solution to the problem.
3) Attainable state changes (legal moves) can be represented in a problem space as transitions from given states to the states directly attainable from them. But considerable moves can also be represented — ie all transitions from one considerable state to another.
4) Any knowledge that the problem-solver can acquire about the problem can be represented in one or more problem spaces.
5) If the actual problem involves acting upon the external world, then the definition of state changes and of the effects on the state of applying any operator reflect with complete accuracy in one or more problem spaces the laws (of nature) that govern the external world.
6) All of these conditions hold in the strong sense that the basic processes postulated require only practicable amounts of computation, and that the information postulated is available to the processes — ie available with the help of only practicable amounts of search.

Simon 1973

In the case of electronics design, it is difficult to see how, without an adequate FSL, either of the requirements of point 1 can be met. If a solution is not completely and precisely stated, then how can criteria for its acceptability be determined? In an industrial context, the final criterion for a successful design is that the customer sees a prototype or breadboard and accepts it as satisfactory. In fact, it is very difficult to progress from the breadboard to the finished product without changes being made and these may have unpredictable effects. The following statement by a layout designer is concerned specifically with PCB design, but the problem produces the same unfortunate effects (or worse) in the case of chips:

That's one area where there's always been a bit of difficulty. The engineer's got a bit of a board there, and he's not particular how much space he takes up. He puts the components down and connects them together and after a bit of fiddling and tweaking he gets it working right. He gives it to the drawing office to lay out and they cram it into the board of whatever size is determined

by the equipment and it doesn't work. There's sometimes a bit of
an inquest, 'Why did it work on the breadboard and not there?':
so all the time there's a struggle to get the breadboard a bit more
like the finished job . . .

Begg 1983a

The criterion argument can also be applied to the specification
itself. How does anyone know when the specification is
complete? The engineer or systems analyst has an intuition
(based on a variety of types of knowledge) of when to stop
specifying, but this is not easy to formalize and automate, for
reasons that will be discussed later.

The second requirement, that there be a definable problem
space, has yet to be met. A problem space for electronics
design would be a context in which there could be a complete
specification of behaviours which the finished machine
should be able to perform (initial state), and a full description
of every electronic machine that it is possible to build which
meets that specification. The latter is at present even more
difficult to specify than the former, since the technology in
which machines are implemented is very volatile and not well
understood at present.

For the third and fourth requirements, Simon's comments
on house design are just as relevant to the problem under
discussion here:

The hopelessness of even trying to sketch the congeries of
elements that might have to be included in the specification of a
problem space proves the greater hopelessness of defining in a
reasonable compass a problem space that could not, at any time
in the problem solving process, find its boundaries breached by
the intrusion of new alternatives.

Simon 1973

Every engineer has encountered this problem at one time or
another. The noble attempts he makes to specify exactly
what it is that the customer wants can be ruined, because of
changing market forces, or new technology becoming
available, or because the customer has changed his mind.

The fifth requirement, concerning the constraints of the
external world, is one in which the greatest efforts are being
made (particularly in engineering). All levels of representation
are underpinned by the physical laws which govern the
behaviour of the material components of a design, as indeed

they must be in order for a successful solution to a particular design problem to be achieved. However, it is commonly thought that if designs can be represented in a sufficiently abstract fashion, implementation details are not important, and that logic design (for example) can be carried out independently of any but the most gross consideration of the materials used for the construction of a design. The separation of the abstract design from the concrete implementation underlies the philosophy of technology independence in design. Adherents of this philosophy believe than an abstract design methodology will insulate designers from a volatile developing technology. Those who deal in FSLs and silicon compilers treat design as if it were as mechanizable as theorem proving, and as Simon comments:

> Condition 5 is always satisfied in theorem proving because there is no external 'real world' to be concerned about.
>
> *Simon 1973*

Parts of the design problem are indeed formalizable and have been automated. Still more areas have been explicitly described as technical knowledge (in the form of text-books, etc). But it will be argued in a later chapter that the engineer still relies on unformalized knowledge to produce the design in its final form. This kind of knowledge is what constitutes his expertise: it enables him to notice and correct inconsistencies and omissions in the final product and to produce rules for generating designs which enable modular design to be carried out with a greater chance of successful recomposition. Similarly:

> . . . part of the architect's skill . . . is embedded in the over-all organisation of his program for design. Part of his professional training and subsequent learning is directed to organising the process in such a way that the major interactions between components will be taken care of.
>
> *Simon 1973*

If effective means of complexity control are used, then Simon's sixth requirement (that there should be resources in terms of information and computational power equal to the problem in hand) becomes less daunting. The sharing out of sections of the problem to different members of the design team implies that each of them will require access to a limited subset of information about the whole design, and that the overall organization of the design will need a different subset, of limited size. The mass of information required by the fine detail of the design is shared out and the smaller amount of organizational detail (eg about interconnections between modules) can be concentrated in a different form.

This analysis of the representability of designs can help to determine the limits of automation of the design task. The first step taken here is to outline a model of the design process, and using this model, to clarify the concepts of abstraction, decomposition and refinement.

Abstraction, decomposition and refinement

Observation of electronics designers at work and of their designs (Begg 1983b) has shown that each subgroup of the design team (the systems designer, the electronics engineer, the draughtsman, etc) uses a separate method of representation for the design. Each time the design was passed from one member of the team to another, it was broken into smaller chunks, which were elaborated in detail. A diagram (see Figure 5.1) representing the stages of the design process in its different forms of representation can be drawn as a hierarchy, or tree structure.

Figure 5.1 *The design hierarchy*

There are many different names used for the levels of the hierarchy: in some contexts they are called *levels of abstraction,* in others, *levels of representation* or *description.* This diversity of names reveals a confusion about the nature and purpose of different kinds of representation.

In design, the complexity of objects to be designed is controlled by considering alternative representations of it, this is called *abstraction.*

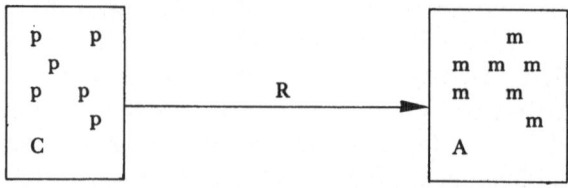

Figure 5.2 *A complex object represented by a simple model*

Abstraction can be modelled as a mapping relation between sets. An abstract view of a complex object (C) is a simplified model (see Figure 5.2): that model (m Σ A, the set of all abstract models of C) is created by application of a set of rules (R) to a set of properties (p Σ C) considered by the designer to be salient to the task in hand. In the case of electronics design, some salient properties of a complex device would be electrical behaviour (giving rise to a model called a circuit diagram) or geometrical structure (giving rise to the models *floorplan* and *layout*).

In the case of geometrical structure, there are two kinds of model based on the same property. They are both created by *structural decomposition.* This process, and its functional relative, can be illustrated in an uncomplicated way by using another complex object, a car. A car is represented in the maintenance manual by several different kinds of drawing. An exploded drawing of the whole car serves to tell the user how the car is put together. One property of the car is selected (the structural contiguity of its parts) to create this model. There are a few interesting points about decomposition, since it is a class of abstractions which produce more or less complex models. First, one can apply the same subset of the rules, r(explode), several times to the parts of a model to get finer levels of detail.

This is known as *nested decomposition*: in the case of the car one element of the exploded diagram of the whole machine can be considered, the stub axle, for example, and an exploded drawing made of that. This comes in handy when that part has to be dismantled and reassembled.

Functional decomposition, however, is not such a simple matter. Take for example the braking system of a car. A model constructed by the method described previously will take the property 'braking behaviour' and represent only the behavioural characteristics of the components of the braking system.

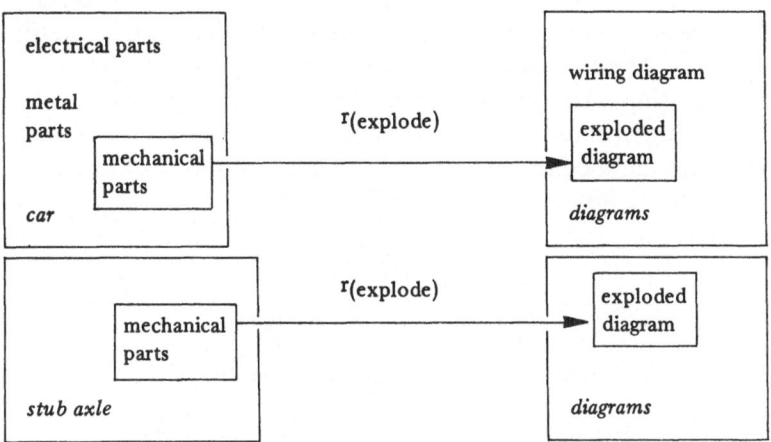

Figure 5.3 *Simple decomposition, exemplified as in the representation of a car*

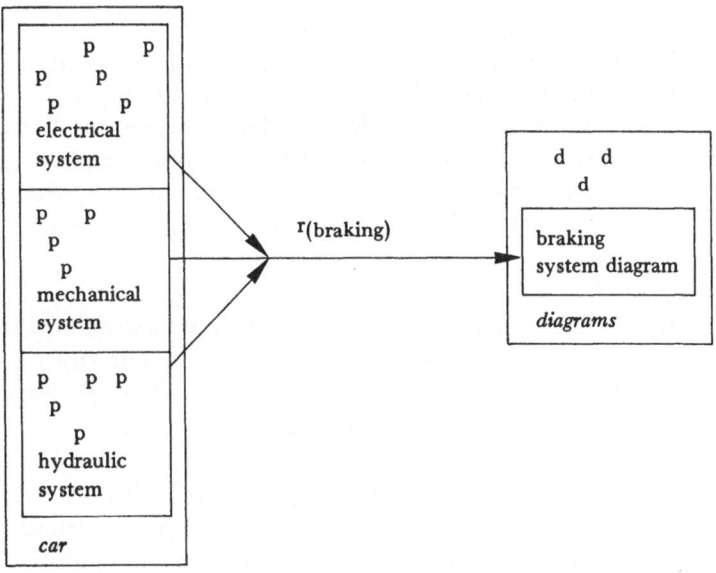

Figure 5.4 *Complex functional decomposition, exemplified as in the representation of a car*

In practice, however, one is not interested only in the behaviour of the parts of a braking system, but in their structure, in order to locate and fix them should they break down. For this reason, a braking system diagram is usually represented as a complex of structures drawn from different functional systems, ie the electrical, mechanical and hydraulic systems. These systems are abstractions on several r(p) relationships. Diagrams produced in this way are often confusing because they conflate several levels of abstraction (functional decomposition and the issue of levels of abstraction will be discussed in more detail later).

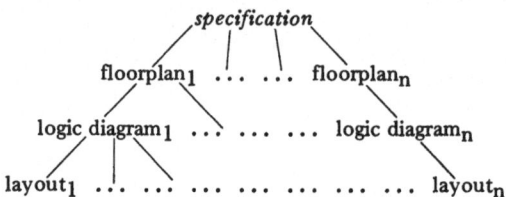

Figure 5.5 *The representation hierarchy*

In order to articulate the various models of the design, many people use the idea of a hierarchical relationship between types of model (see Figure 5.5). The use of hierarchical structures to model complex objects and processes is paralleled in the development of the structured programming movement in software engineering (Dahl 1972), in related techniques in AI (Sacerdoti 1977), and in the structured design movement in VLSI design (Mead & Conway 1979).

Mead and Conway used a model of design based on hierarchical structuring of levels of description in which each level 'contains and controls' the level beneath it. The top level of such a structure is considered to be the most abstract, and this is the specification of the machine's behaviour.

It is particularly important to represent the specification of the machine explicitly, completely and verifiably, since it is the point of reference for the rest of the design, as well as the starting point. Formal specifications are recommended (eg by Rem 1981) as a means of ensuring the integrity of the design. Not only is the formally stated design amenable to proof, like a theorem, but it is in a form of representation clear enough to be accurately translated by machine into lower levels of representation (as in silicon compilers). The design process can (theoretically) be represented in this way as a deterministic algorithm. When an informal representation of the specification, or any of the other levels of the design, is used, it is not possible to verify the mapping between representations, and the transformation process is vulnerable to undetectable errors.

However, the models used by the various members of a design team can hardly be considered to be formal models in this sense. If one accepts the need for increasing formality in representations as a prerequisite for increased automation of the design task, then the salient characteristics of each model must be determined before an adequate formal model can be built. In the case of the VLSI chip, it is possible to isolate four kinds of representation (see Figure 5.6), and say that they model different aspects of the device to be manufactured.

representation used	property modelled	basic unit of model
formal specification	I/O behaviour	functions
floorplan	gross physical structure	blocks
logic diagram	logic function	gates
layout	fine physical structure	polygons

Figure 5.6 *VLSI abstractions*

It will be plain to engineers with some experience of electronics design that this characterization is too simple. In the process of generating a model, the representation used constrains the result in more complicated ways than the decomposition relationship described. In order to see why, it is necessary to consider how design is carried out.

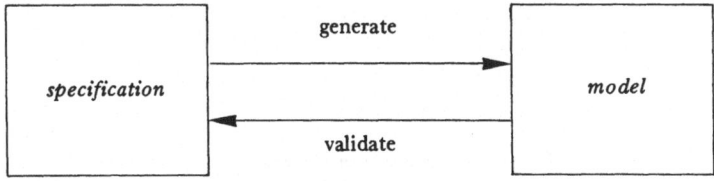

Figure 5.7 *The design cycle*

The design process can be characterized as a cycle of stages in which a new model of the design or a part of the design is generated at each stage. This model is checked for integrity against some higher level specification. The products of each stage can then be considered as a specification for the next. The final testing of a product is just another stage of validation of a model, this time a concrete one, of the designer's idea.

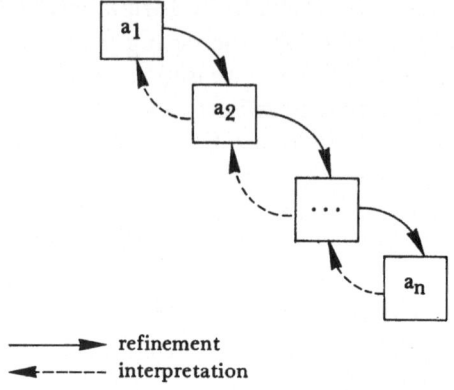

refinement
interpretation

Figure 5.8 *The hierarchical model of design*

Design can thus be viewed as a set of transformations from 'abstract to concrete' and vice versa. In the process of producing a specification, the designer is forced to abstract, or generalize, over related areas of the design in an effort to produce a comprehensible description of a complex object. The transformation down the vertical dimension of the representation hierarchy (see Figure 5.9) is often thought of as the inverse of abstraction.

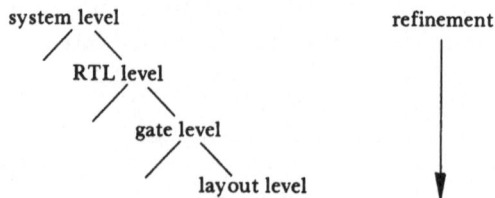

Figure 5.9 *The representation hierarchy in digital design*

However, there are several questions to be asked about this hierarchical model of design. What is the refinement process? How does it differ, if at all, from decomposition? How can interpretation between models, and validation, be made?

Refinement is the principled transformation of one model of a device to be manufactured to another. In order to formalize the design process, the transformation rules required to effect refinement should be rigorously specified, but in human engineering practice they are a more or less informal collection of heuristics.

Refinement can be distinguished from decomposition only if the idea of a level of abstraction is reconsidered. In this model of the design process, the boundary between two levels of abstraction is considered to be crossed when the designer changes models. It is necessary to do this when a new aspect of the design is considered (see Figure 5.10).

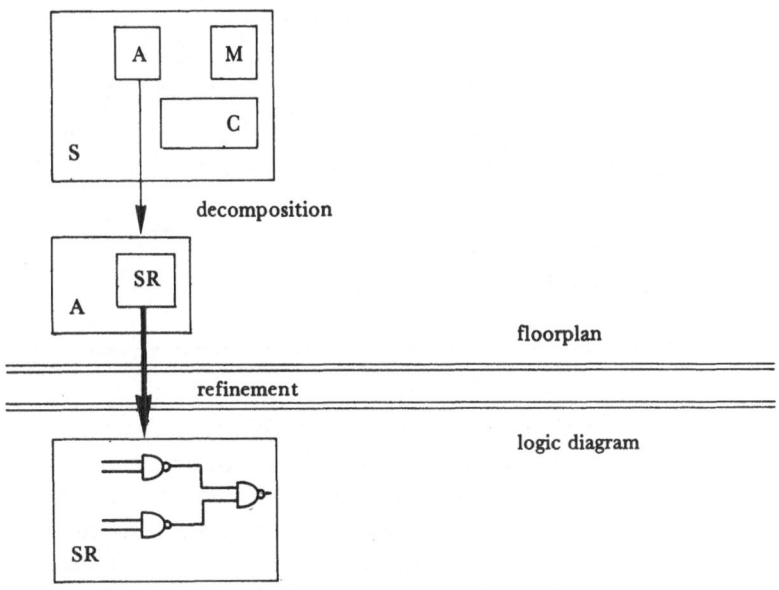

Figure 5.10 *The boundary of two models*

In this case, as long as the concern of the designer remains in the topological relationship of blocks, decomposition can continue *ad infinitum*. However, as the level of detail increases, it becomes more and more difficult to consider other aspects of the design, since the complexity of the model increases. In order to preserve the usefulness of the model, the designer usually stops decomposing the floorplan at a point where it becomes feasible to refine an individual block into logic diagram form and think in terms of gates. Note that in order to use the block as a specification, the designer has to refer to its original *functional* definition as generated by the specification.

This difference between decomposition and refinement defines two types of abstraction: (a) as a level of detail (determined by grain size) and (b) as a level of analysis (determined by the type of model used). The rule sets used for these transformations (see Figure 5.11) differ in that, for decomposition, rules can be applied recursively *ad libitum* and are reversible (ie can be recomposed without further transformation), whereas those for refinement have only the former property. There is a many to one mapping between models which are not recomposeable, ie there is no sure way of working from logic diagram back to an identical floorplan, since structural information is sacrificed from the logic abstraction to make way for information about function.

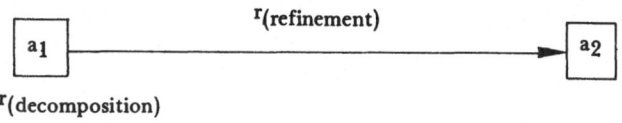

Figure 5.11 *Transformations between and within models*

Each successive model used in design is derived from a different set of rules and a different subset of properties of the original complex conceptual object. It can thus only be considered *vis-à-vis* other models by interpretation. Hence, validation of one model against another becomes problematic unless the interpretation rules are explicit and clear.

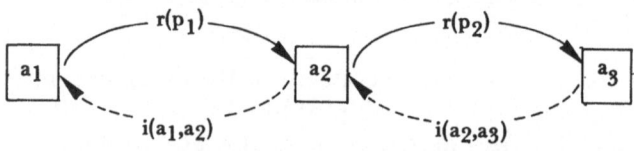

Figure 5.12 *Generation and validation of models*

Figure 5.12 illustrates how refinement rules (r) use properties

(p) to generate each successive model (a), whereas interpretation rules (i) define semantic relations between the models.

Some implications of this model of design can now be considered. Most designers use the forms of representation which evolved during the development of their craft, and those forms of representation do not lend themselves easily to formalization. Hence, it is difficult to see how there can be principled relations of any sort between these diverse models. If such an analysis is true, the technique of hierarchical decomposition cannot produce verifiable designs unless the forms of representation used to model successive levels of abstraction are considerably revised.

Concepts of formal and informal knowledge

The structure of the model reveals a potential for infinite recursion if care is not taken to establish stop and start conditions. The iterative nature of the design cycle requires reference to an arbiter at both the top and the bottom of the cycle. This is usually a human decision guided by various types of informal knowledge (see Figure 5.13).

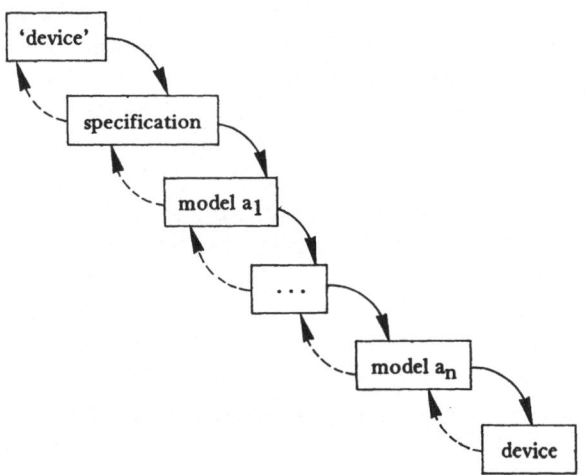

Figure 5.13 *A sequential model of design*

The first step (which will be discussed later) is the production of a specification from some ill-defined conceptual device. The final step, producing a real device, is usually referred to as *implementation.* If one accepts the notion that design consists of a sequential series of transformations of abstract models into increasingly concrete models, then refinement and implementation can be considered to be identical processes. Likewise, one can consider each model to be acting, in some sense, as a specification for the next.

There are several problems associated with the model in this form, which are shared by the hierarchical model: if each model is made very 'pure', excluding all concerns except those of its own level of abstraction, then logically, the only way of ensuring a correct design will be to wait until the device is manufactured and tested. In practice, there are many rules governing representation (eg design rules and drawing standards) designed to avoid specific classes of errors only describable outside the current level of abstraction. However, the invention of representations which consider such constraints are not guaranteed ideal, by either the criteria of humans or computers.

Moreover, although the description of function is theoretically best isolated from the description of structure (cf. Jones 1980), in practice, structural components can be used as shorthand for functional units, as a 'simplifying abstraction'. A logic designer may include elements of circuit and layout descriptions in the logic diagram, eg a chip, representing a complex logic function. This all makes the formal description of the design *process* very difficult. The design *documentation* can be depicted as a neat hierarchy but the actual process of design has far less clear-cut boundaries.

In a large design team, it is easy to imagine that the parts of the team responsible for each stage of the design can be, and in fact are, independent, each using knowledge and information of which the others are necessarily ignorant. For the systems designer, the system diagram is the completed design. For the draughtsman, the layout is the completed design. However, information and advice does flow across these barriers, and there are other sources of information and advice apart from the design team itself.

The operations to be carried out at each stage are of the same abstract sort, being refinements from functional to structural descriptions of finer implementation details. However, they require several different sorts of knowledge to make them possible. A project manager, deciding on how to break down the system specification into submodules, will have considered factors ranging from the purely practical to the highly technical, but he will not use the same detailed knowledge of a machine's micro-structure that a draughtsman uses when deciding where to place a particular IC on a board, or cell on a chip. The systems designer consults with the potential end users and with the rest of the company that employs him. He uses information about marketing strategy, cost limits and resource allocation. The draughtsman uses knowledge of fine details of technological limitations in drawing a practical layout, using a better model of the physical and electrical properties of the medium than that embodied in the design rules. Despite the relative informality of the various levels of representation, these heuristics relate the otherwise independent levels of representation and allow the design process to proceed. A variety of expert knowledge, experience and intuition is used in directing design decisions at any one level, in addition to the sort of technical facts about function at that level.

What is the difference between these sorts of knowledge? Three classes have been described: (a) technical facts, (b) information from below the currently used level of abstraction, encoded in design rules or implicit in the design engineer's expertise, and (c) information from elsewhere, which is not included in any of the representations used because it is non-technical. Here the concept of technical needs to be clarified. For present purposes, technical means 'within the explicit and formal knowledge of the electronics engineer': technical knowledge is the sort one finds in textbooks and the sort which can be encoded into algorithms for use as CAD tools. Non-technical knowledge is that which lies outside the formalized set of abstraction dimensions and outside the domain of the currently used set of models of the design.

Each domain of technical knowledge has a well-defined boundary, set by the form of the representation. Only well-known objects and relations can be represented. Everything that can be included in a level of abstraction is considered a technical fact or rule. Thus, each level of abstraction can be considered as constituting a sub-domain of expertise. The rules for generating the representation encode part of the explicit knowledge about the domain, and there is usually a set of well-known procedures which can be performed within it. Design tools and design automation programs which automatically translate or generate designs use these well-known algorithms.

However, design automation projects often fail to deliver the kind of speed and efficiency demanded by the industry and are limited in range and scope. In formalizing the process of design, too many non-technical facts and rules are left out which contribute, in human performance, to good design practice. For example, in automatic gate-packaging programs, the designer has an overview of the whole board and the pattern that the components form on it. The program only has the netlist to work with: a wholly inadequate representation of the packaging problem since it excludes knowledge of the semantics of the nodes and links which model the circuit. In addition, the majority of CAD programs consider only the constraints on design that propagate downwards from the specification. Human designers often take into account the constraints propagating upwards from levels below that at which they are working. For example, a logic designer may include in the logic diagram strong hints, either explicit or implicit, about the siting of particular components.

To understand how implicit knowledge about design can be drawn out and included in CAD systems, both the representations and the sets of rules used to create models must be reconsidered. The idea of a specification has become more and more abstract in academic circles, until only a set of functional equations, or logical propositions can qualify as a formal specification. The formal specification *per se* is a valuable abstraction, but it cannot contain the grab-bag of expert and non-technical knowledge needed to make a complete enough description of a device for the manufacturer.

The premise of complexity control is that by constructing abstract models which consider aspects of a complex object and which are salient to its design, testing and manufacture, it is possible to construct a concrete implementation which is, in some way, the composition of all the abstract models, or at least, of the properties manifest in the original conceptual device. The idea of 'device' is explored by the designer with respect not only to the behaviour of the device but also to concerns, such as environmental ones (Who will use the device? Where will it be used? How long will it be used for?, etc) and to resource constraints (such as how much time can be spent in design, whether such a device already exists, how much the customer is prepared to pay for the device, the pre-existence of a range of implementation technologies, etc). At present properties like these are eased into the design process informally, in the informal specification.

Moreover, the initial specification and subsequent models have to be revisable if it is discovered that further refinement or implementation is impossible. This is yet another argument for explicit rules of interpretation, so that backtracking and revision of models can be made. The sequential model of design illustrated in Figure 5.14 can only be verified by implementation: if a parallel rather than serial model is constructed (see Figure 5.14), some of the problems become tractable.

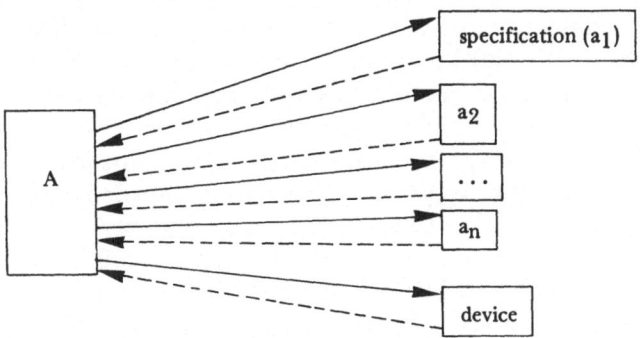

Figure 5.14 *A parallel model of design*

The technique of complexity control, as described here, is to build successively more concrete models of a complex object, each of which is created by considering the object from the point of view of a limited subset of its properties. In the parallel model, it is possible to imagine keeping track of the properties modelled so far in the design, and making a record of their relationships. This model of the relationship of models (A) can be considered as a reference point for validation of the design at any level.

The function of an abstract model, in science or engineering, is to help us understand or explain a complex phenomenon. The complexity of the electronic systems design problem has yielded a little to the techniques of abstraction, but there is still a long way to go. An over-complex model is not only difficult to understand, but is an admission of ignorance. If it is not possible to produce a suitable abstract model for an object, related to others which are already understood, then it cannot be said that it is understood at all. Further, an over-simple model is not necessarily the best: the salience of a model depends on the use to which it is put. One can happily model the orbits of the planets as perfect circles if one has no need to navigate the solar system. If design automation is to be achieved, large tracts of informal knowledge about electronics must be captured in the form of computer software. Before this can be done, many more details of design expertise must be made explicit, and currently used models must be rationalized and related to each other.

The methods of complexity control through increased use of formalisms and abstract models described here are advocated by mathematicians and practised mainly by software engineers. How does this theoretical view of design fit in with the observed practice of electronics engineering design? No designer ever starts from scratch: few start from a formal specification. This chapter has also shown how knowledge about design is encapsulated in different forms of representation. Other ways of encapsulating knowledge about design were described in Chapter 3, eg how designers try to compile their knowledge of successful designs in cell libraries and in design methods. Further, from the description of design tools in Chapter 3, it is clear that where explicit rules and strategies can be formulated, attempts have been made to mechanize aspects of the design process (eg in automatic layout routines, silicon compilers, etc) with some success. In Chapter 6, the possibilities for further mechanization, both of technical and non-technical design strategies, will be considered.

Techniques for electronics engineering from artificial intelligence

> I would have liked to call this paper 'Can a Eunuch discuss Sex?'.
> We are all eunuchs as far as interaction is concerned. Very few
> people have experienced real interaction with a computer.
>
> *Nick Negroponte (in Guedj 1975)*

Over the last ten years the development of software
technology has advanced as quickly as hardware technology.
Research in AI has begun to produce ideas for working
intelligent systems, and has contributed many new approaches
to improving the human-computer interface. How some of
these new techniques can be applied to CAD will be discussed
here.

Most AI work on electronics design has been concerned
specifically with LSI and VLSI circuits. It is a new area in
which design methodologies are both being evolved and
actively designed, and one in which the underlying technology
has been sufficiently volatile over the last ten years to
destroy most of the preconceptions that designers had about
what its future would be. Old methods of design have been
thoroughly exposed and tested and new methods have been
used on a scale borne out of desperation of ever being able to
keep up with the pace of development. Moreover, that pace
was being held back, not only by the limitations of design
methodologies but by the secrecy and security precautions
which shroud the design and production of commercial semiconductors. These are not conditions which make for rapid
evolution of new methods. In the past, the development of
scientific ideas and the new technologies they support has
benefited from the sharing of knowledge. This sharing, at the
very least, avoids the necessity of having different groups of
people reinventing the wheel.

The evolution of design methods

One of the problems created by rapidly developing technologies is that of the revision of habitual working methods in a way that is acceptable to designers already working in the field:

> . . . How can you take methods that are new, methods that are not in common use and therefore considered unsound methods, and turn them into sound methods? In other words, how can you cause the cultural integration of the new methods, so that the average designer feels comfortable using the methods, considers such usage to be part of their normal duties, and works hard to correctly use the methods. Such cultural integration requires a major shift in technical viewpoints by many, many designers. Changes in design practices usually require changes in the social organisation in which the designer functions . . .
>
> When methods are new and still considered unsound, it is usually impossible in traditional environments to recruit and organise the large numbers of participants required for rapid, thorough exploration and for cultural integration. Therefore, new design methods normally evolve via rather *ad hoc*, undirected processes of cultural diffusion through dispersed, loosely connected groups of practitioners, over relatively long periods of time.
>
> *Conway 1981*

The importance of the professional context of the individual designer and the environment of the design team as a whole to the effective modelling of the human-computer interface has already been pointed out. Historically, design methodologies have evolved over centuries rather than decades, so the idea of inventing new design methodologies is new to us. Nevertheless, one of the factors which must be taken into consideration in redesigning the human-computer interface for CAD is how to increase its acceptability to the designers who use it.

It is possible to design a VLSI circuit equivalent to a whole system, such as a minicomputer. This has brought about radical changes in design decomposition methods and design representations. An example of such change is the KBVLSI project at Stanford and Xerox PARC. The project is concerned with the design of an expert system (PALLADIO) for digital systems design based on a four-fold abstraction scheme different in many respects from those observed by the author. Stefik *et al.* (1982) and Conway *et al.* (1982) described in their own terms the difference between functional decomposition and abstraction. Functional decomposition consists of splitting up designs into a *component hierarchy* corresponding to functional modules. Abstraction is a way of stating design problems in a way that allows certain critical issues to be considered early on and across the full breadth of the design process.

> Abstract solutions are descriptions that stand for an equivalence class of detailed solutions.
>
> *Stefik* et al. *1982*

The abstract solutions to which Stefik and Conway refer are represented in what they call *description levels.*

> [Table A] summarises four experimental description levels that we are developing. Each description level has a set of terms that are composed to form systems and a set of composition rules that define legal combinations of the terms. The concerns of each level are characterised by specific classes of bugs that can be avoided when the composition rules are followed. Each level has a behavioural meaning as well as a structural meaning so that descriptions are also runnable.

description level	concerns	terms	composition rules	bugs avoided
linked module abstraction (LMA)	event sequencing	modules forks joins buffers	token conservation fork/join rules	deadlock data not ready
clocked registers and logic (CRL)	two-phase clocking	stages register transfer transfer functions	connection of stages	mixed clocked bugs unclocked feedback
clocked primitive switches (CPS)	digital behaviour	pull-ups pull-downs pass transistors	connection of switch networks ratio rules	charge sharing switching levels
layout	physical dimensions	coloured rectangles	lambda rules	spacing errors

Table A *Description levels in PALLADIO (from Stefik* et al. *1982)*

Each level of description represents different aspects of knowledge about the design, and composition rules are designed specifically to avoid common faults. Thus preemptive design decisions can be made, which prevent buggy designs even being thought of. Design can be viewed as a search through a space of possible machines that can be generated dynamically according to known specification constraints. Composition rules represent various parameters that can be used to prune the search tree early on to speed up selection of an appropriate design.

Stefik *et al.* pointed out that there are differences in their approach from conventional CAD tools, at the structural as well as at the implementation level. It is not just that they use heuristic programming techniques:

> The philosophy behind our approach differs significantly from that used in the construction of silicon compilers. In a silicon compiler, the desired behaviour of a system is specified in a language at a single level. The compiler converts this behavioural description into a structural description in a standard format. This fails to exploit many possibilities in the design space. In our use of multiple descriptions, each level specifies both behavioural and structural information. User-chosen transformations can be made in the design at many different levels all the way down. For example, in the Linked Modules Abstraction level, one might find optimisations which yielded substantially different structures before deciding whether to use two-phased clocking or self-timed circuits.
>
> The use of multiple levels for describing hardware has been tried many times. For example there are logic descriptions and register transfer descriptions (van Cleemput 1979). We believe that the logic descriptions are too isolated and the RTLs are incomplete and insufficiently formalised . . . In essence, those descriptions were not designed for synthesis. They provide no composition rules, optimisation rules or bug characterisations. Our goal is to understand and formalise descriptive levels whose utility derives from their coverage of critical design concerns.
>
> *Stefik* et al. *1982*

This approach differs from the conventional silicon compiler approach in that it is an integrated, general-purpose design system with a radically different way of representing designs. It offers a redrawing of the levels of representation which are commonly used, and suggests that the inclusion of composition rules and description of characteristic bugs captures sources of knowledge not previously formalized. However, one problem that was not foreseen was that of adequate and accurate communication and translation between levels of representation. This was the rock on which the KBVLSI project foundered. (The project is at present 'on the back burner'.)

Recasting the representation of the whole design task in new forms and ignoring successful formalizations of specific areas creates more problems than it solves. Implementation of rule-based systems in place of existing, well-formalized but fragmentary algorithms may lead to a situation in which the way toward a well-formalized and simple solution is obscured by a mass of poorly understood processes. Designers of CAD might well be advised to concentrate on well-structured, formalizable problems: but there is no guarantee that a good computational theory which deals with the problem of complexity control will ever be evolved. How do we proceed in the meantime?

Formal and informal decision-making procedures

One thing that can be done is to separate out what can be performed by the 'handle-turning' of formal methods from that which is not yet formalized, or that which is as yet unformalizable. Sussman (1978) noted 'the importance of having knowledge which tells us how to avoid the need for algebraic manipulation as opposed to more powerful manipulators.'

> Believing in the ultimate power of mathematical manipulation is one of the most common difficulties encountered by students learning electrical circuit analysis. Students often grind out 'impossible' algebra in the course of solving a homework problem, even though a little thought will reveal an algebraically feasible approach which depends on a small insight into the network being analysed.
>
> *Sussman 1978*

Some of these insights can be represented formally. There are new types of formal representations for design problems arising out of the work of mathematicians. The conception of systems design as the manipulation of a network has lead to the application of more and more sophisticated branches of graph theory to the problem (Rosenfeld 1974, Claus *et al.* 1979), and the idea of hierarchical decomposition to the use of notions from set theory through category theory (Chen & Mead 1983; Rem *et al.* 1983). The idea of increasing the speed of computation through parallel architectures has created interest in formal representations of parallel processes

(Milner 1980). Fundamental to all these efforts is the notion that a good notation is an essential aid to the control of complexity, in that it acts as a reliable shorthand for the expression of complex ideas.

> A simple analogy may help to motivate the adoption of formal notation. The reader will recall the following kind of problem:
>
> > John has three red toys and James has more green toys than the total number owned by John and Jack together. Jack has ....
> > How many toys are owned by each boy?
>
> The way to handle such problems is to reduce them to symbols and then manipulate algebraic expressions. One feature of such problems is that a certain amount of irrelevant information is usually included. In the analogous situation with specifications, there is normally plenty of irrelevant information; unfortunately one cannot rely on all the essential information being present. It is to be expected that specifications will also be more tractable in a mathematical notation.
>
> *Jones 1980*

The idea of a good notation in mathematics is close to that of a good representation in AI. In order to provide good representations of problems in mathematics or AI, one has to have a very clear insight into the structure of the problem in question. In some areas, this structure is not at all clear:

> The term representation is used clearly (almost technically) in AI and computer science. In contrast the term *knowledge* is used informally, despite its prevalence in such phrases as *knowledge engineering* and *knowledge sources*. It seems to be a way of referring to whatever it is a representation has. If a system has (and can use) a data structure which can be said to represent something (an object, a procedure, whatever ...), then the system itself can also be said to have knowledge, namely the knowledge embodied in that representation about that thing.
>
> *Newell 1980*

Formal methods have been devised for dealing with some of the sub-problems in electronics design, but many important aspects of complexity controlling knowledge about the process of design have not yet been formalized, simply because the design process itself is undergoing evolution and the structures which could embody the relevant knowledge are not yet available to the designer (in Conway's terms, not yet 'culturally integrated'). Some design methods have been devised whereby the underlying technology is manipulated in order to enable a simple formalism to be used to generate the design (see Rem 1983). However, this avoids one set of problems only to create another set at a later point in the process.

It was pointed out in Chapter 5 that although design may look like an uncomputable problem, if it is properly divided it can be at least partially solved. Much of the design automation work done since 1970 has been concerned with the accurate computation of aspects of design and whole designs for limited purposes. If an adequate decomposition strategy and a good representation for a particular design problem can be decided on, then refining it can be made automatic. However, some areas of the design problem, notably that of partitioning the problem according to knowledge of a variety of problem structures at different levels of abstraction, remain intractable by traditional algorithmic methods. It is these areas which will respond to heuristic programming techniques. After all, algorithms and heuristics can be considered as closely related means of representing knowledge:

> Algorithms are only heuristics which are so powerful that we can make guarantees about their use.
>
> *Lenat 1983*

At this point it would be useful to consider what the words *knowledge* and *decide* mean for the purposes of CAD systems design. Knowledge in the ideal CAD system (according to the criteria described in Chapter 4) will be a representation of any mechanism that allows design decisions to be made. Design decisions are the choices a designer makes between any of several alternative implementations of a design at successively lower levels of decomposition and refinement. Design decisions must be judged according to a specification of the behaviour of the finished machine. Some familiar problems crop up again here. First, in Chapter 5 it was argued that it is impossible to fully specify a machine, so the criteria for judging the optimality of a design will always be incomplete. If there is not a complete set of criteria by which to judge design decisions, how can a choice be made between alternative machines?

This will always be a problem in a system which is designed

to include both formal and informal decision-making procedures. Both procedures are necessary, since some of the design process, such as specification, is difficult to formalize. Some knowledge about design can be and has been implemented directly as an algorithm. Some may be fundamentally intractable. Newell (1980), in his paper on the *knowledge level*, points out the problems of formulating knowledge: one cannot consider knowledge as a complete formal entity (such as the logical closure of a set of axioms) because there are simply too many counter-examples which show that people cannot possibly know all the implications of such a formal statement. He concludes that knowledge-level models are not determinate, but are only an approximation, ie knowledge is 'radically incomplete':

> The term *radical* is used to indicate that entire ranges of behaviour may not be describable at the knowledge level, but only in terms of systems at a lower level (namely, the symbolic level).
>
> *Newell 1980*

In other words, knowledge may be implicit in the structure of lower levels of abstraction and this knowledge will be inaccessible to explicit formalization.

This conclusion sounds depressing, but is in some senses equivalent to the problem of full specification discussed earlier. In the absence of a complete set of relevant data (a common situation) the designer resorts to informal methods, rules of thumb, which work, 99 times out of 100. He uses his own knowledge about the way that each level of abstraction is typically structured given a particular context, and works from there, rather than from complete specifications, or complete, explicit knowledge sources. Different groups of people have different aims in the pursuit of knowledge which leads them to value one kind over another. This is the source of the continual friction between academics and engineers, scientists and technologists. It should only be a source of conflict in CAD systems design in so far as it is not possible to decide in which category (formal or informal knowledge) to place a particular aspect of the design task.

The problem-solving approach
Several methods have been devised by AI researchers for representing knowledge about electronics design. At the

Massachusetts Institute of Technology, Drew McDermott's DESI program used a problem-solving approach with a single-level predicate calculus description language. The design task was reduced to a set of essential design actions:

1 DESIGN
2 MAKE
3 Constrain
  CONSTRAIN
  SELECT-VALUE
4 Change
  FIX quantity
  BIAS
  COUPLE

which were performed on a limited range of primitive device types with no internal structure.

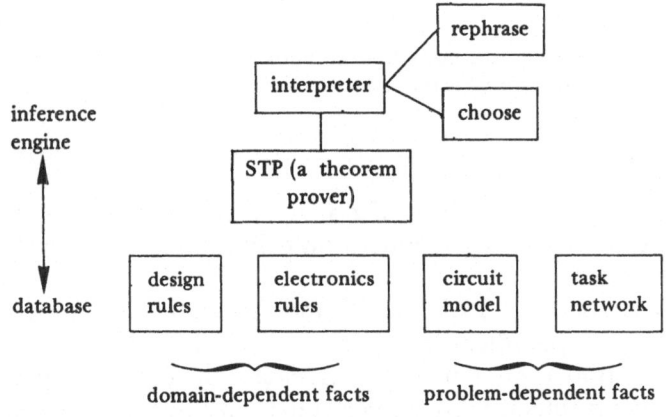

Figure 6.1 *Structure of the DESI program (McDermott 1978)*

The essential mechanism of the program was a manipulation of the problem description, in which the REPHRASE section of the program 'exploded' elements of the task to be performed into recognizable sub-tasks until they matched specifications of partial solutions already 'known' to the program. McDermott summarizes it thus:

> . . . This theory provides for the following stages in design: transformation of the given design problem into a form that matches the specifications of an indexed partial solution, followed by solution of the sub-problems so created. This involves

71

acquisition, connection and constraint of components and their control attributes.

McDermott 1978

Actions to be taken in design were deduced by the theorem prover and woven into a task network, representing a plan of action whose elements could be further decomposed. If more than one plan was deduced, the CHOOSE section of the program looks for sub-tasks, or asks the user for rules about how to make choices about plans for solving sub-tasks. One sort of rule in this section was RULE-TOGETHER, which created a more general plan out of two or more old plans. This constituted a primitive type of general composition rule, and could be the basis of a program which acquired new rules for design. McDermott concluded that what was needed was a better theory of the acquisition and nature of composition rules:

> . . . I have strong intuitions that composing partial solutions in a domain like electronics is guided by knowledge about composition in the abstract.

McDermott 1978

One of the major problems with DESI was how to represent this abstract knowledge about composition in a single level of representation.

Contrast between PALLADIO and this earlier system from a different AI paradigm shows that this single level of representation approach fails to elucidate the complexity of the design problem. Predicate calculus can form the basis of a good specification language, but its properties do not express the action of composition very well. This is because they are designed to capture concepts and relations in a textual format which forms an unsuitable basis for thinking about physical relations. To a mathematician, the issue of representation at this level is irrelevant. There is no reason why relations between the shapes of things should not be represented as a set of logical propositions. However, if one level of representation is used to describe more than one level of abstraction in a CAD system, confusion is bound to result, and with that confusion, inefficient calculation.

De Kleer and Sussman's SLICES program (Sussman 1978) uses a rule-based language and makes use of analysis by propagation of constraints (proposed in a seminal paper by De Kleer *et al.* 1977) to assign component values in a circuit. A more principled decomposition method is used, since only selected parts of the circuit are considered at one time, following the practice of expert circuit designers, who 'use terminal equivalence and power arguments to reduce the apparent synergy of the circuit so that their computational power can be focused'. By a technique which Sussman terms 'knowing the form of the answer' the circuit is effectively decomposed into a set of relatively independent sub-problems which can be represented in an abstract fashion. Knowledge about the form of circuits is encoded in the decomposition method, but again, the single level of representation chosen leads to the splitting of the design into a large number of primitive sub-solutions which then need to be associated by some more general composition rule.

In considering how to go about CAD systems design, one of the fundamental problems raised by the idea of an intelligent system is to decide whether to provide the designer with a tool for design, or the manufacturer with an automatic designer. It was argued earlier that neither complete formal specification of desired system behaviour, nor complete formal representation of knowledge required for design is possible. Yet the proceedings of over 20 annual conferences have been involved with attempts to achieve design automation. What techniques have been used to create automated design aids for VLSI?

The components described overleaf constitute a representative selection of those in use in design automation software. They can be used either separately, in a conventional interactive CAD system, or together, as a silicon compiler.

Design automation
There are many types of design language, working at different levels of abstraction. Two examples of successful design languages are: *MacPitts* (Siskind *et al.* 1982), a specification language which compiles into mask level specifications in Caltech Intermediate Format (CIF) (another design language, which has become a standard for mask level descriptions), and *occam* (Taylor & Wilson 1982), a language for describing parallel architectures. The aim of a design language is to describe the behaviour and/or structure of a circuit without ambiguity.

73

An *extractor* is usually a fairly simple translation program which specifies the connectivity of a circuit from its behavioural description (see Newell & Fitzpatrick 1982).

A *data path generator* is a program which constructs data paths from standard cells. This entails the use of massive computations sometimes ameliorated by the use of heuristics to constrain some of the more potentially explosive searches (see Shrobe 1982 — 'a design aid that is tailored to a limited domain and which embeds a lot of knowledge about the particular features of that domain can greatly reduce the complexity of designing in that domain'). Shrobe's data path generator uses standard cells to produce symmetrical arrays.

A *control section generator:* the control section of a chip is a more complicated entity than a data path and requires correspondingly more complex tools — it is practically an automated designer in miniature. Anceau's IRENE uses a:

> style selector: an expert program using techniques of Artificial Intelligence which will implement the strategy of the selection of the control section organisation from the design experience of the research group.
>
> Moore-Mealey transformer: provides two different means of decision-making for the components of the control section; Moore — switching between several candidate control sequences and Mealey — using conditional generation of commands; which is used depends on style selector. Components are repetitive blocks of ROM, PLA.
>
> generators of repetitive components: which also generate their logical descriptions.
>
> PLA optimiser: a topological optimiser or compaction algorithm.
>
> a topological evaluator: which uses statistical methods to measure the relative deformability of a block of a floor-plan and its potential connectivity.
>
> layout assembler: PASCAL-like language for program which composes cells. Because the electrical properties of all the cells are known and described, then this composition automatically provides an electrical simulation.
>
> *Anceau 1983*

The silicon compiler approach represents one end of a dimension in CAD which runs from the 'fully automated' system to the 'designer's toolbench' (see Friedenson 1982), where a selection of CAD tools can be used in a looser relationship with one another via a front-end processor which concatenates programs and translates data from one format to another. The advantages of a fully automatic system is that human-introduced errors are less frequent, and the ease with which data can be produced for simulation and verification purposes.

> The use of algorithmic level design specifications and the automation of the refinement process will change many aspects of VLSI design in addition to the layout task. It is common practice to analyse a design in an attempt to remove any apparent flaws and to guarantee a circuit's correctness prior to fabrication . . . Generating designs using MacPitts will alleviate much of the verification task. Layouts generated by MacPitts are correct by synthesis, obviating the need for design rule checking . . . As new techniques for VLSI design are developed, these can be incorporated into MacPitts. Possibilities include automatic test vector generation and generation of testable, self-testing and fault-tolerant circuits. Automation of such techniques would be difficult if layouts continue to be generated by hand. These possibilities are conceivable only when designs are specified at a high enough level of abstraction and are refined to the layout level automatically . . .
>
> *Siskind et al. 1982*

> The VLSI designer is also a programmer. He designs ultraconcurrent programs and he, (like the conventional programmer) should not be confronted with the details of the physical representation. The layout in the plane must be done automatically, by a silicon compiler, without interference or consultation of the programmer . . .
>
> We will not consider it a defeat of our method if our layout algorithm would require a special fabrication process for the realisation of our programs as VLSI circuits. It could, for example, be that we would like our VLSI circuits to have one or two additional layers of metal. We are willing to complicate the fabrication process if that is the price we have to pay for allowing the programmer to abstract from the physical representation of his products.
>
> *Rem 1981*

Rem's concept of design driven technology is a departure from the usual position which requires that fabrication constraints should be built into the system and actually constrain the process of design. But this position is a necessary one for design automation. If systems are to be completely automatic, they must be more formal. In order to achieve this simplicity, formality and rigour in the design stage, sacrifices may have to be made at the manufacturing end of the design process.

Building intelligence into design procedures

In the description of McDermott's and Sussman's problem-solving approach to design, the idea of using constraints on design to guide the design process was introduced. Design constraints can propagate both ways in the abstraction hierarchy, downwards from the specification and also upwards from the limitations of the implementation technology. In most current CAD applications, design rules are a coarse representation of the underlying technology, because current design rule-checking programs lack intelligence.

A more subtle version of the current design rules can be suggested. The fabrication of microprocessor chips is a multi-disciplinary affair: to a greater extent even than the design process, involving chemists, physicists, mechanical engineers, production engineers, and experts on various kinds of measurement and testing where equipment is at the known limits of technology. Though some engineers think it unnecessary to take even the most gross consideration of the physical nature of the components they design, the people responsible for fabrication are very aware of the repercussions that the subtleties of the manufacturing process have on the feasibility of various design strategies, both from the point of view of physical viability and cost per chip.

It would be unrealistic to expect designers to cram their heads with details of manufacturing processes, but the creation of design rules flexible enough for different applications would be an improvement on the rigidity of the current system. One can imagine more sophisticated design rule generators for which each etching, deposition or implantation layer was considered in turn, and rules were determined for that mask tailored to the amount of growth, shrinkage or undercutting occurring at each level. These parameters could be made flexible (ie changeable), able to respond to changes in technology or the needs of particular applications. For example, the expense of creating a large, full custom design makes it worthwhile to create a more subtle set of design rules in order to reduce designs to their very minimum size. With a smaller, less complex design, one could afford to use rules which were less efficient in saving space or improving speed. This sort of detail could of course only be feasible in the context of an automated design system, in which the layout was actually generated in accordance with design rules. A draughtsman would have difficulty in learning such a complex rule set. At first sight, a comprehensive program like this might be thought to be too computationally expensive, but it may be possible to use the decomposition hierarchy of the design itself to speed up rule checking (cf. Johnson 1982).

THE COMPUTER AS A CONSULTANT

Another approach to the design of complex computer aids comes from the emerging expert systems paradigm. Weiss *et al.* (1982) have described an expert system that acts as a human-computer interface for a suite of complex oil-well log analysis programs. Based on the earlier example of SACON (Bennett & Englemore 1979), an expert system which advised on the use of a big structural analysis program, it goes further than its predecessor, in that the 'consultant' and the program on which it gives advice are interactive. The user never actually has direct communication with the underlying algorithm:

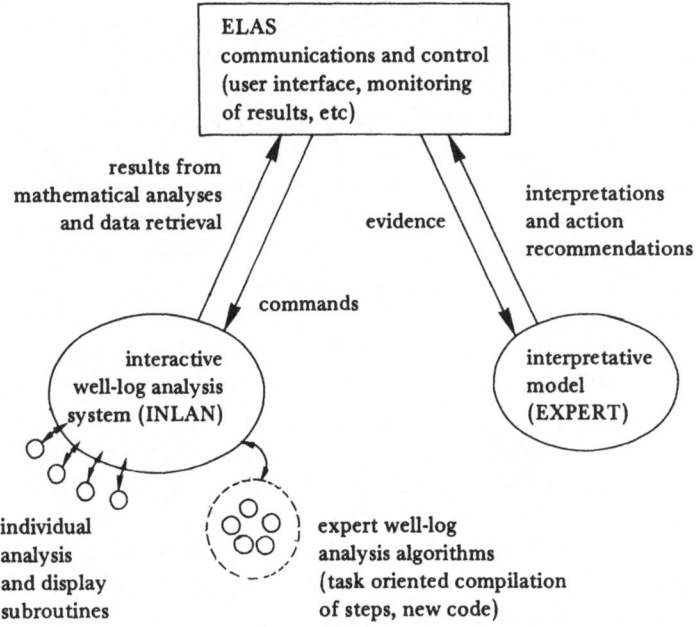

Figure 6.2 *Overview of ELAS communication (from Weiss et al. 1982)*

In order to have lines of communication between programs as well as between human and computer, program transparency must be sacrificed. Inter-program communication must be looked at very carefully to ensure that what the interface programs tell us is being done *is* actually being done. One opinion is that since engineers are working in higher level languages than in the past, they have no need to know the inner workings of the program, any more than a software engineer needs to know how to read a core dump.

In the case of very complex programs it may be as well to maintain the opaqueness of the system for the sake of the operator's mental health. Complexity control can also be used at the level of CAD systems design. To insulate the designer from the complexity of the software tools, interface programs (themselves) must be allowed to take decisions so that the designer is presented with relevant information in a simplified form. This requires a knowledge-based interface of some sophistication.

Other advantages of an opaque expert consultant system for CAD are that the interactive explanation facilities that can be provided with an expert system are in themselves an aid to simulation and debugging of designs. Although an interface program would prevent the designer from communicating directly with deep levels of the system, it would provide a means of presenting an acceptable picture of exactly what the CAD tools in use are doing. Such a system could also provide automatic knowledge elicitation as part of the interface so that the informal design methods being evolved by the designer could be made explicit, evaluated and possibly incorporated into the system.

The expert systems paradigm has provided a whole range of potentially useful tools to enhance the human-computer interface, but what is more important, its use as a controller of complex programs provides a way of using formalism and conventional algorithms in a CAD system while keeping them separate from (and incorruptible by) heuristic procedures representing informal knowledge about design.

Returning to Figure 4.4, and bearing in mind what has been said about expert systems, the levels of a CAD system can be modified as shown in Figure 6.3.

This scheme serves to separate programs which perform well-known and well-formalized functions over sub-problems in design from those which are intended to capture and use less well-formalized knowledge over the whole design. Representations of the various levels of abstraction of the design can be said (in Newell's sense) to reside below the knowledge level, although the representations themselves may embody important knowledge.

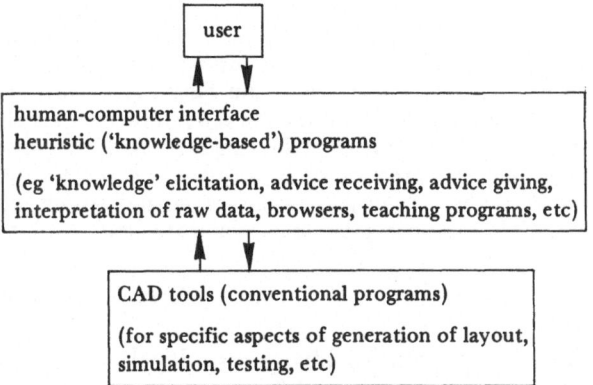

Figure 6.3 *Communication with CAD software via an expert system*

CHOOSING LANGUAGES FOR DESIGN

Having proposed a means of controlling the tools which operate on the design, it would be useful to look again at the way designs are represented. Stefik and Conway's PALLADIO system uses a fixed set of abstraction levels with a different description language for each one. Sussman's SLICES program composed sections of a circuit from primitives using only one type of representation. Both approaches have draw-backs: it is difficult to ensure correct translation between levels if more than one description language is used, but one representation will not be adequate to capture all the relevant aspects of a specific level of abstraction.

Chen & Mead (1983) offer a multi-level simulator in which the user can define within generous limits the level of representation at which he wishes to work. This is achieved by using a dual hierarchy of syntax (for ease of specification) and semantics (for functional abstraction), and the flexibility given by an embedded language within the general purpose programming language.

It has been observed in IC layout languages that an embedded language — a language supporting graphics primitives in an existing programming language — has the generality and flexibility in the specification of designs that an interactive graphics system usually lacks. The effort of making a graphic system as powerful as an embedded language is essentially that of supporting a general purpose programming language. It is much more sensible to let the compiler of an existing language do the work. The same philosophy applies to a specification language for simulation. We build into a programming language the simulation algorithm and an interactive user interface ... for testing the design. One specifies cells in an embedded simulation language by invoking primitives for transistors, nodes, syntactic cells and semantic cells. These primitives are pre-defined in the language. With the power of a general purpose programming language, users can then specify functional abstractions, data abstractions and various data types at any level according to their conceptualisation of the design.

Chen & Mead 1983

This technique of representing abstraction levels seems to provide the flexibility and accessibility required by the system criteria outlined earlier. It should not be too difficult to apply such a technique to design rather than simulation. Design and simulation differ, however, in that with a simulation the appearance of the finished circuit is known. Ideally, a design environment should provide the means to consider alternative designs at all levels.

GUIDING SEARCH THROUGH A DESIGN LIBRARY

Bobrow & Goldstein (1980) have suggested a solution to the problem of representing alternative versions of a (software) design at one time, and allowing alternative recompositions. This is done with a *context-structured database*. A frame-based representation is used with a *context principle* which ensures that when one retrieves the values of attributes of a node, one does so in a particular context in which only the values assigned in that context are visible. A series of *layers* can be built up which are isolated from each other, so different designers can work together on parts of the same design, yet their partial designs can be merged in different ways.

The designs can be accessed using a variant of the Smalltalk 'browser', which has an hierarchical structure in four levels (category of class, class, method within class and method), but which proves unsuitable when combined with the context-dependent structure of the database. There is a need for more research into the problem of complexity, as many researchers feel that hierarchical structures are inadequate. They also note problems in the human-computer interface which arise from such a complex program: the power of the system has to be presented in a useable form, according to the cognitive requirements of the designer. Both the gross structure and the fine structure of such a complex design aid must be well attuned to the designer's needs, and often these are not known.

The concept of structured design is based on the hierarchy as an appropriate representation of designs for the purposes of complexity control. Mead & Conway (1980) base their model of the design process on a hierarchical relationship of 'containment and control' existing between levels of abstraction. However, if one starts to consider the possibility of user-defined levels of abstraction, or if one wishes to explore the possibilities of composing various partial designs, then hierarchical structure of this type becomes a hindrance rather than a help. What is needed is something more flexible than a strict hierarchy, perhaps along the lines of a lattice (see Jones 1980) or a similar concept of partial ordering (Mitchell 1978). If the various levels of representation are considered as a set of algebras, then the idea of an equalizer, from category theory, suggests a translation set which can transform a representation at one level into a representation at another without the need for a strict ordering relationship.

One way in which some of these problems can be approached is to make use of a meta-level knowledge base, which encodes knowledge about knowledge. Davis & Buchanan (1977) define meta-level knowledge as that which enables a program 'not only to make use of its knowledge directly, but may also be able to examine it, abstract it, reason about it, or direct its application'. They categorize four types of meta-level knowledge:

knowledge about	*is encoded in*
representation of objects	schemata
representation of functions	function templates
inference rules	rule models
reasoning strategies	meta-rules

The concept of meta-knowledge may be a way to avoid the

rigidity of always having to apply hierarchical structure to the abstraction scheme. Some of the more well-formalized and better understood aspects of a design are best represented in a hierarchal relationship, but it is useful also to be able to treat all abstraction levels together as one large domain, with each separate representation considered as a sub-domain. A meta-level representation of the correspondences between these sub-domains would act as a key to translating one sub-domain into the terms of another. Reasoning about appropriate representations will be necessary, too, in an environment in which user-defined descriptions are made possible.

So far, the requirements for improved CAD systems design include several complex concepts: multiple levels of representation of a design, facilities for meta-level reasoning about design strategy, etc. The structure that controls this system will have to be extraordinarily powerful and versatile. A candidate structure is the blackboard, a feature of the HEARSAY-II speech-understanding system (Erman *et al.* 1980), which has been adapted for use in several other AI applications. This architecture is described by Hayes-Roth, from whom the following description is quoted:

> The blackboard architecture has four definitive elements: a) entries, which are intermediate results generated during problem solving; b) knowledge sources, which are independent, event-driven processes that produce entries; c) the blackboard, which is a structured, global data-base that mediates knowledge source interactions and organises entries; and d) an intelligent control mechanism, which decides if and when particular knowledge sources should generate entries and record them on the blackboard. Partial solution 'islands' emerge, one entry at a time, in different structural partitions of the blackboard. New 'solution islands' appear and existing islands grow wherever the opportunities are most promising. Eventually, mutually supportive partial solutions merge to form a complete solution.
>
> *Hayes-Roth 1983*

In the context of a CAD system, entries on the blackboard would be design units (blocks, gates, components, etc) suggested as refinements of a particular level of representation. *Knowledge sources,* in the present terms, are sets of production rules which govern the search for or design of design units, and their combination into partial designs. Examples of such rules might be relevant to one level of representation or concern design data from more than one level.

> ... knowledge sources frequently transform entries at one level of abstraction into entries at another level, some knowledge sources ... operate bottom-up. They aggregate several lower level entries into a smaller number of higher level entries. Other knowledge sources operate top-down ... (Yet others) operate within a single blackboard level or between different blackboard 'panels' (subdivisions of a single level of representation). Thus, the blackboard architecture can combine knowledge sources embodying different inference mechanisms in a single problem-solving system.
>
> *Hayes-Roth 1983*

The blackboard itself serves a unique purpose in separating the different kinds of knowledge source from each other, while allowing them communication and influence over each other via blackboard entries. The blackboard also organizes all partial and complete solutions generated for a particular problem by marking the relationships of the entries to one another.

The control mechanism which governs blackboard activities is basically a scheduler which looks at a list of 'knowledge source activation records' to decide which action to perform next. Triggering a knowledge source does not lead to immediate activation of the rule's right-hand side. This allows the control mechanism to examine the range of possibilities for action and take a truly global view of the activities of the system.

The blackboard structure offers features which are invaluable to the design of intelligent CAD tools. It provides unlimited access to design data through its multidimensional representation scheme. In addition, it keeps track of different versions of partial designs and coordinates the activation of various sources of knowledge through a meta-level control mechanism.

Some AI techniques have been described here which could perform the functions required of a good CAD system. However all the possibilities have still not been explored. Implementation details of the application of many of the AI techniques mentioned would be difficult to specify at this stage, and guarantees about the performance of most of the systems could probably not be given. Hopefully the general points about the structure of an ideal CAD system (based on the criteria of Chapter 4) are taken. In Chapter 7 the design of such an ideal CAD system will be discussed, and the techniques described here will be linked together to suggest the possibilities of a working system.

CHAPTER 7

An 'ideal' CAD system

At this point, so many models, criteria and requirements have been presented that the reader is likely to be confused. The design task has been described (Chapters 3 and 5), requirements for a better CAD system have been discussed (Chapter 4), and some of the new approaches to interactive and automatic design systems have been considered (Chapter 6). Having got this far, the question 'What makes a good CAD system?' looks more, rather than less, complex. One way to simplify a complex problem is to decide on a clear goal: here the aim will be to sketch the outline of a specific CAD system which fits the requirements of designers.

Components of a system
To summarize, the following requirements were outlined in Chapter 4: an ideal CAD system should provide:

1) automatic generation of design data compatible between design tools;
2) high-level design languages and tools;
3) a supportive and adaptive human-computer interface;
4) access to information in the system, combined with security against accidental damage or loss of files;
5) a dedicated, customized system which provides good, relevant information quickly.

This allows a few of the functional components of the system to be isolated. (Following the advice given in Chapter 5, this system design will proceed in a 'top-down' fashion, specifying the required behaviours of the system first and systematically refining them later.) The first point indicates that some kind of *translation* facility will be needed to produce design data in a useful and valid form. The supportive human-computer interface which provides access (with security) to different levels of the design can be packaged as a *consultant* function. The consultant concept will be discussed fully later. Good relevant information can only be stored in some sort of a *library*.

Continuing onward, other requirements can be drawn from the user profile described in Chapter 4:

1) the system should be easily understood by hardware engineers;
2) all members of the design team should have easy access to the required design information;
3) there should be an increase over the current speed of information retrieval, verification and simulation (the non-creative elements of the system);
4) the human-computer interface should provide graphical I/O wherever possible.

These points indicate that some kind of natural-language like interface (based on a subset of English and engineering jargon) is needed, as well as a very good interactive graphics editor. A browser function (see Chapter 6) would be useful in searching design databases.

In addition to these basic points, the CAD systems designer should bear in mind that:

1) the complexity of the design task is high and should not be made more intractable by an over-determined command structure;
2) the users of the system will be both casual and regular, naif and sophisticated, in any permutation;
3) to use a user's model of the task which is too detailed and accurate may perpetuate unwanted features of the manual methods of draughting and design.

Two separate issues emerge here. First, in order to cope with the different demands of the design team, the system control structure should be designed to cope with multiple levels of interaction (while still keeping track of the development of the design). Second, the designer needs feedback from these multiple levels which is easily recognizable, so as to allow early debugging of designs. Some sort of simulation of designs at all levels of representation will allow this to be done.

These first thoughts summarize the overall behaviour of the system, but there are some additional points which must be considered before starting to refine a system design.

1) What is the system needed for? Different applications and technologies require different tools. There may be a level of abstraction above which operations can be considered to be technology independent, but low-level constraints must also be accommodated.

2) How much should the CAD user be allowed to 'interfere' with processes which are automatable? Is interaction really necessary? The silicon compiler camp is firmly against this, but given the volatility of the technology and that a truly technology independent representation of electronics design is probably difficult to achieve, is it not sensible to use human 'wetware' where it is most efficient, ie in creating new answers to new questions, quickly.

3) How much can the unaided machine achieve? How well can human knowledge be put into mechanical form? Should the system be able to produce rigorously correct designs for a limited range of applications automatically, or should it provide a flexible range of tools to be used by designers working in any application?

These three points can only be resolved if the designer refers to the existing markets and at which of these the CAD system is to be aimed. There is no one correct solution for any design problem, and the design of CAD systems is no exception. The idiosyncracies of individual designers and of different design teams will be sufficiently disparate that there will be a place for most types of CAD system. Given the inherent complexity of the task, any system ought to cater for a broad range of applications, including IC and PCB design in digital design. The designer should be allowed to interact with the CAD system at all levels so that automatically generated elements of designs can be optimized manually when necessary.

Basic structure and operation
Having defined some of the behaviours which would be expected of this ideal system (called CADCONS for CAD CONSultant from this point), it is possible to go a little

further and outline its high-level structures and operations. Fundamental to this design exercise is the idea that the design of CADCONS should reflect the task structure and classes of expertise that the author's study showed to exist in practice. Hence, the objects and operations specified for this system are analogous to the forms of representation and sub-tasks in the design process carried out by the systems designer, electronics engineer and draughtsman. In Chapters 3 – 5 the design process was described as having three main constituents:

1) specification, done by a systems designer;
2) translation and refinement into a variety of alternate representations, done by an electronics engineer; and
3) layout, done by a draughtsman.

However, it was pointed out in Chapter 5 that both the engineer and the draughtsman perform similar kinds of translation task, but at different levels of abstraction. Separating out the actual *operations* carried out at the different levels of the organizational hierarchy, leaves specification, decomposition, translation/refinement and recomposition. Specification remains a unique function because of the informal relationship between customer and designer. Although specification could be thought of as a kind of translation, it needs to be kept separate because of the necessity of human interaction at this point.

In CADCONS, these operations are modelled by the system functions SPECIFY, GENERATE, SIMULATE and REFINE. This is a variety of hypothesis testing: a design is hypothesized, modelled and tested, and the hypothesis modified according to the results of testing. Specification and refinement can be done automatically at all but the topmost level of representation (ie specification), and in all cases the functions are performed in an iterative sequence.

For the time being, design objects can be thought of as existing at an indeterminate number of levels of representation between top (specification) and bottom (layout) of the dimensions of abstraction/decomposition (see Chapter 5), since different representations are required for different applications, and designers have personal preferences about which they will use. Within one level of representation, decomposition of the whole design results in a set of modules, or partial designs. These can be decomposed in turn and eventually recomposed to form the completed design. Difficulties will obviously arise in striking a balance between providing tailor-made modelling tools for each level of abstraction, and standardizing representations so that successful recomposition can take place. This is a problem that can only be resolved by negotiation between the CAD system designer and the target design team.

In order to explain the structure of CADCONS, performance at one level (specification) will be used as an example to demonstrate how all the levels work. Specification level is for several reasons not the easiest example, not least being that there are few extant computer aids for specification. However, a demand exists for such aids, and the top level of CADCONS represents an attempt to show how a high-level design tool could be constructed.

A sketch of a computer aid for specification

Clear specification of behavioural requirements is essential to the quick production of good quality, verifiable designs. Since engineers are resistant to learning how to use design languages and specification languages, facilities will be provided in this CADCONS which will allow a more human dialogue to take place between designer and computer. The designer will specify the behavioural requirements of the design in a natural-language like interface (NLLI) (see Figure 7.3), and the computer will translate this into a suitable specification language. An interface of this kind can be found in the software design environment PIE (Bobrow & Goldstein 1980). This formal specification, in its internal representation, will act as a reference point for automatic generation of additional refinements of the design. Most importantly, it will act as input data for a very high-level simulation program so that the designer can check his progress at each iteration of the 'generate, simulate, refine' loop.

The prime object in the system is the formal specification (SPEC), which is produced by compilation in a functional unit called GENERATOR, from data provided by a dialogue between the user and a CONSULTANT program (see Figure 7.1).

Decomposition of the design proceeds by taking the initial specification and transforming it (in a GENERATOR program) to an alternative representation which can then be presented to the user in a simulation and adapted or refined where necessary (see Figure 7.2). If the specification is stated formally, it makes the job of writing the translation program (essentially a compiler) very much more easy and the results are much more reliable.

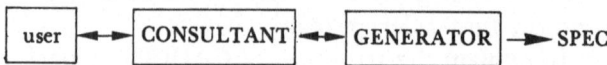

Figure 7.1 *Production of an initial specification*

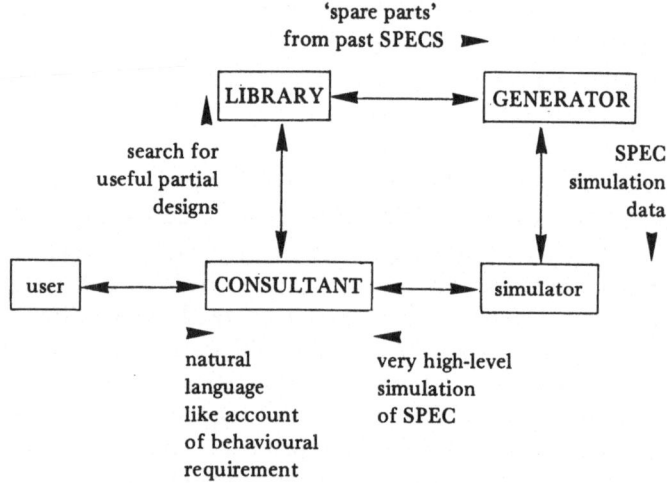

Figure 7.2 *Refining a specification with the aid of CADCONS*

Partial designs (from parts of specifications to individual components at the layout level) which have proved reliable will, wherever possible, be re-used so as to speed up the design process. Few designs will need to be started from scratch. The part of the system that makes this possible is the LIBRARY, which holds specific kinds of ready-made parts for use in different categories of machine and in operations at different levels of abstraction.

Classification in the different sections of the LIBRARY is by entity (named module or component: it is possible to summon a 'block of memory' of a particular size and shape, or a component such as a 'shifter' or 'barrel shifter' at a different level of abstraction) or by category (class of module or component: it is possible to review all available blocks of memory or all types of shifter or even types of barrel shifter or, at a lower level, all standard cells that have named I/O characteristics and are of a certain size). The user can access the LIBRARY directly by using the BROWSER facilities of the CONSULTANT (see Figure 7.4).

Recomposition of the specification is a matter of linking together old and new partial designs which fulfil the user's requirements. It is difficult to say at present what sort of composition (or decomposition) rules could be created to deal with this process, but further research is planned in this area (Begg 1982). One of the tools to be included in CONSULTANT is a knowledge acquisition system (KAS), which will help to capture these rules. The KAS uses the NLLI to enter new rules, facts and definitions into the specification knowledge base (KB) in the course of the dialogue with the user. (The model taken for this KAS is from Reboh 1981.) At other levels, where composition rules are better known, composition algorithms exist, which when necessary, can be called from an algorithm library (AL) (see Figure 7.3).

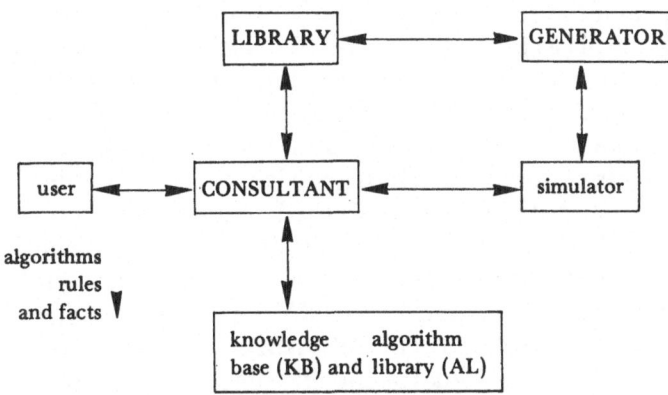

Figure 7.3 *Acquiring a variety of kinds of knowledge*

The KB and AL are located together in Figure 7.3 to emphasize the function of CONSULTANT as an expert system that reasons about appropriate computations. The algorithms in AL are seen as extended right-hand side (action) parts of rules. This enables well-known, tried and tested

91

design tools to be incorporated into a rule-based system.

In the process of specification, it is anticipated that the problem will be keeping user interaction at a high level. The specification language should deal with a level above a register transfer language, where functions only are discussed (ie the language used will be one which produces a functional equation). Information about low-level constraints tends to emerge as the specification is made and, in Chapter 5, arguments were made for the capture of this sort of data. CONSULTANT will contain a layered cache (part of BLACK-BOARD, cf. Chapter 6) in which to keep such constraints separate for later use when lower levels are activated.

The CONSULTANT interface

Some of the components of CONSULTANT have already been mentioned. A NLLI allows the KAS to accept definitions of new concepts and rules (see Figure 7.4). It is a command language consisting of a restricted subset of English, with a limited ability to guess at meanings of incorrectly spelled words or sentences with syntax errors. A similar language is used in the program design system PIE (Bobrow & Goldstein 1980). Associated with NLLI in CONSULTANT will be an interactive graphics editor which can be used in the normal way. An interesting line of research would be to discover as many means as possible for CONSULTANT to build up a picture of the habits, preferences and competence of the design team through NLLI. This information could then be used to modify the sorts of information an individual designer is exposed to, offering advice when it is needed, for example, and suppressing unwanted information.

In CADCONS the KAS also acts as a librarian, organizing all the stores of design information and making sure that input to the KB, AL and LIBRARY is never paradoxical, redundant or outdated. The BROWSER allows the user or CONSULTANT to look for specific objects or classes of object in LIBRARY. Browsers are now (1984) fairly common utilities in the software of some American made work-stations. In order to protect the user from the change in command language caused by a change in system level, the user has access to BROWSER only through NLLI. CONSULTANT allows controlled access to all levels of design information, while cushioning the user from the complexities of file systems, editors, etc.

92

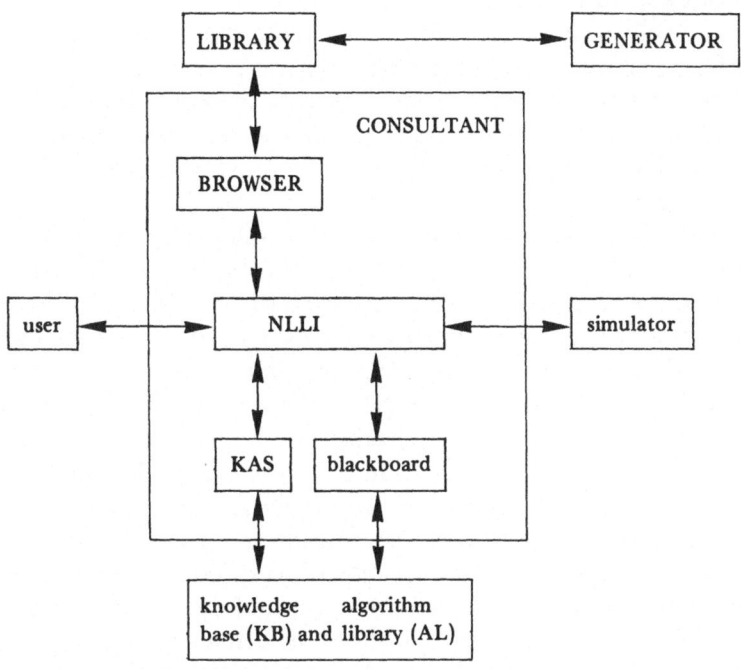

Figure 7.4 *The components of CONSULTANT*

The main purpose of CONSULTANT, however, is to act as a controller for the whole system. There are many levels to the system and many knowledge sources to tap for each level. The structure that best fulfils the requirements of monitoring and controlling a multi-level system is that of a blackboard (see the description of HEARSAY-II in Chapter 6). The blackboard structure is layered, so that the development of different versions of a design can be traced and different refinements of partial designs can be compared. The blackboard provides *meta-level* control of the system, monitoring the use of design tools and ensuring that the relevant information is being provided from an appropriate source.

Amongst the points that CONSULTANT will be watching for are: Where are the big searches? Which user or what program knows the heuristics to control them? Where are the big computations? Is there any way of reducing their size or importance, or of eliminating them altogether by using other methods? These are rules about using other rules, reasoning about appropriate computations.

As well as gathering information from the designer and offering simple searches through libraries, the system should be able to use its various rule bases to give advice, when requested, and to explain why particular actions are being taken. This facility may be one of the most useful for debugging designs.

The blackboard control structure enables the knowledge bases pertaining to each level of representation to be called into action when that level is accessed by a designer. In this way, the designer can refer to earlier stages of the design and to versions of partial designs that may have been discarded at that time. The name CADCONS perhaps underestimates the power of this system: its operation can be conceptualized as the collaboration of a 'community of experts'; it is more like a CAD consultancy firm than an individual consultant.

The multiple levels of representation mentioned earlier are only useable because they are 'clean', verified, automatically generated design data which are generated by a suite of cross-compiler programs, similar to those used in a silicon compiler, which can transform the original formal specification data into lower level representations. Wherever possible, GENERATOR uses spare parts (partial designs from the LIBRARY) in order to work up the refined representations, thus saving as much time and effort as possible. The designer can obtain access to all the representations of the design produced by GENERATOR, and modify them by using an interactive graphics editor in the usual way. However, all additions and modifications to the design are 'cleaned up' by GENERATOR before being displayed, so as to prevent the introduction of more obvious errors. Serious design faults (in structure, rather than in detail) will always be difficult to detect and correct, but the use of feedback (to the designer) at every level is designed to show up structural faults at an early stage of the design cycle.

If the designer feels that an application is well known enough to specify correctly and fully first time, then GENERATOR can be allowed to proceed unchecked, acting as a silicon compiler.

The display of design data output at each level of abstraction is enlivened by a simulation showing how well each partial design works. Simulation of whole designs is only performed effectively when the simulation data have been generated automatically. Hence, even when freehand manual designs have been constructed, they must be rationalized by the CAD system before simulation. SIMULATOR takes data from GENERATOR as input and applies appropriate simulation algorithms from the algorithm library. Raw simulator output is processed according to the designer's requirements before it is re-presented in a comprehensive form.

Advantages of an intelligent CAD system

In summary, three sorts of operation are performed by this system: specification, abstraction and recomposition. Specification involves a negotiation between the user and the computer about what constitutes a viable machine. The user's concept might be more or less formal, but the computer's has to be formal and verifiable for the rest of the system to work properly. Starting with a high-level formal specification, this system adds to the accessibility of the levels of compilation that go on in a silicon compiler, by allowing the user to intervene, in a controlled way, with the design during the process of automatic generation. Depending on the sorts of representation appropriate to the application and the relative compatibility of these representations, the user might be able to contribute to the design at six or seven successive stages if this was required. As much use as possible is made of past designs in the generation of new abstract models. Recomposition is a more complex process than perhaps it might seem. If it is done automatically, there is an advantage in that simulation data are generated in one step of the design process. The simulation of the final, recomposed, design can be subjected to tests produced by routines in the algorithm library.

There are three properties of this system that make it especially desirable:

1) the re-use of proven elements of past designs;
2) the combination of conventional software and intelligent, knowledge-based systems using well-known, tested procedures where possible; and
3) the provision of means for controlled user interaction at multiple levels of representation.

The design of the system is intended to provide a flexible

set of design tools which can work together automatically, but which the user can manipulate in cases where a special Gestalt view of a design is needed. The tailoring of the system to particular applications is particularly important to the shape of specific instances of the system: some elements, eg libraries and knowledge bases, will be drastically different from application to application (perhaps even from site to site) not only in their contents but also in their structure. A library constructed to hold information about components of PCB designs will be indexed on entirely different parameters from a library of standard cells for VLSI applications. The construction of the system allows the coordination of knowledge about electronics design from many different sources, including from the designer.

Proposals for systems of this sort have been and are being considered by CAD systems makers throughout the world of high technology. This particular proposal is designed primarily to demonstrate two points: first, the potential impact of techniques from AI on all fields of CAD, and second, how this impact does not eliminate the human designer from the loop, but extends his capacity to cope with the design of complex objects. It is an exercise in futurology, in so far that the solutions it provides are yet to be implemented, but the problems they solve are already with us.

CHAPTER 8

Summary

CAD is one of the most exciting fields of research in a discipline which has come to be known as cognitive engineering. It lies at an overlap point between psychology and mathematics, AI and engineering, where the theoretical and the arcane come to have practical applications. This book has described its application to the problems of electronics design, but CAAD (computer-aided architectural design), CAE (computer-aided engineering) and CAM (computer-aided manufacture) are all fast growing application areas. The advances that can be made in the forcing house of industrial applications may well contribute to our general knowledge about how to use and design computers with AI.

It can be argued that in order even to understand what the term 'artificial intelligence' implies, some knowledge of human cognitive psychology must underly that of computer science. Models have been presented here which show something of the nature of the design task and the designer, along with results of analysis and classification of problems encountered in design. This kind of essentially psychological knowledge helps to identify the kind of new techniques which will solve previously intractable problems in CAD.

The task model on which the 'ideal' CADCONS system is based (see Chapters 5 and 7) characterizes design as a process of translation or transformation from one form of representation to another. Each new form of representation embodies a different kind of information about a design. This model differs from that imposed on the designer by the organizational hierarchy of the workplace, and it was chosen after consideration of the similarity, at a high level of analysis, of the tasks performed by each member of the design team. Having obtained the task model it is possible to discover which parts of the process are already automated in CAD tools and to match some of the remaining operations to programming techniques from AI.

One of the main aims of the author's studies has been to isolate criteria for a 'good' CAD system. Some of the results obtained (see Chapter 4) were similar to those reported in the literature on the human-computer interface and would suggest themselves to common sense (the need for a supportive, crash-proof environment, for example). Though the majority of these requirements are well known, software designers still persist in neglecting them. Other requirements, specific to the needs of electronics designers, were also discovered and are discussed below.

The second group of results were concerned with classifying problems in the domain of CAD systems and the social context of their users, the design team. This was necessary so that soluble problems could be distinguished from (practically) insoluble ones. The whole environment of the design team was observed, taking into consideration the reasons why a CAD system was bought in the first place and what it was expected to achieve, noting the relative ages, educational history and experience of the users, etc. In this way CAD use was set in a context in which the values used to produce judgements about what makes a good system are more evident (eg management want CAD to boost production, staff want CAD to make their jobs less tedious). This is interesting in itself in terms of group dynamics, but more importantly, it enables the CAD systems designer to identify classes of problems that the design task poses.

Before these results can be applied it is necessary to separate problems that can be remedied by improvements to the CAD system and those which are problems in 'social engineering'. In the latter case the influence of, for example, market forces on the CAD manufacturers, or the fear of automation on the CAD users, create problems which manipulation of the mechanical parts of the system cannot ameliorate (see Chapters 3 and 4). The class of remediable faults can be separated into those which are due to bad programming practice (where poor feedback of design information is given to the user, or where systems are poorly integrated) and those which require new techniques for their solution (eg incorporation of ill-structured knowledge in rule bases, provision of advice through a consultant program, etc). Analysis of the design task reveals the influence of the social organization of the design team on the way the design task is partitioned. A relationship can be seen between the numbers involved in the design team and methods of decomposition: the smaller the design team, the less obvious the demarcation between levels of representation used by team members. The number and variety of levels of representation is obviously affected by task complexity, too, since decomposition is a method of complexity control, but the explicitly required use of particular forms of representation by subgroups of a large design team has more to do with professional subcultures and organizational hierarchies than functional necessity. Particular forms of representation, like design languages, used by different design teams vary by organization rather than according to a different kind of training or education: although it has been the case, especially in VLSI design, that an individual moving to a new company can heavily influence the design style of the new design team.

Subgroups of a design team also have favourite schematic representations, eg the flowchart or block diagram of the systems designer, the logic diagrams, Petri nets and state tables of the electronics engineer, which embody facets of knowledge about electronics and which are useful at particular stages of design. However, observations of small (two man) design teams and individuals working on their own, show that the connection between stage of design and level of representation is not rigid.

The third class of results is concerned with matching technical solutions to practical problems. One of the most important questions here was: What can be automated and what can't? This led to the consideration of some deep issues of formalism and representation. Design has long been considered an ill-structured problem for which it is difficult to 'mechanize' in software. One of the first findings of research into AI was the importance of using the correct representation for the problem. This makes solving the problem possible, in the same way that a good notation in mathematics makes thinking about mathematical concepts easier.

It is difficult to find new representations which will enable the mechanization of additional areas of the design task. The obviously mechanizable parts have already been incorporated into CAD tools. In the remaining ill-structured areas, knowledge about design remains implicit, either because the area is not well known (the influence of the rapidly changing technology is particularly strong) or because it does not fit into the representations which are habitually used. It is difficult in some cases to know what an appropriate new representation would look like.

In addition, the analysis of levels of representation discussed in Chapter 5 shows that as well as excluding certain types of knowledge, representations used in electronics design are conceived of and used in a more flexible way than the top-down direction recommended by the approved methods of structured design. Several different kinds of abstraction were noted, used for different purposes in the design. This observation has led to the recommendation of a multi-dimensional approach to representing design at different stages of completion.

In order to combine the known methods for automating design (established CAD algorithms) with heuristics which capture informal knowledge, the 'expert consultant' approach to a knowledge-based system was proposed. In the CADCONS system, described in Chapter 7, existing procedures for CAD are considered as extra large action parts of production rules. The inclusion of a blackboard in the control structure of this sort of system should enable the usually opaque silicon compiler technology to be open to user interaction at all levels of representation. Moreover, the blackboard architecture enables one to use and control upward-propagating constraints, thus giving the designer the opportunity of doing with the machine what is done in practical design: feeding back information about difficulties in implementation details to 'higher' levels of representation, which can then be modified.

Looking so carefully at the kinds of problems that CAD sets the systems designer has naturally led to the identification of areas about which not enough is known. How can algorithmic and heuristic knowledge best be combined? How does one elicit the knowledge needed for building computer aids for specification, for example? What kind of formal language should be used as an internal representation? The language chosen must have characteristics which allow it to be used as a reference point for checking the progress of the rest of the design: meeting the specification is the only criterion possible for a successful design. If this criterion is not made available to the system, automating any part of the design is a hit-or-miss affair.

What kind of input is possible to such a specification aid? In this study it has been indicated that engineers resist using specially constructed design languages. The languages have nice formal properties which allow designs to be verified, but they are difficult for engineers to learn, since the medium of statements (in what is effectively a programming language) is very different from the language of graphics which they habitually use in developing a design. In the CADCONS system, a command language based on the terms used by the designer is recommended, but this is by no means the only solution.

The question which underlies these issues is: How can appropriate forms of representation be found for capturing what may be purely procedural (but is in any case not often put into words) knowledge from experts of any kind? It has been pointed out that forms of representation embody different kinds of knowledge: in the construction of expert systems it is thought that particular kinds of problem are suited to particular problem-solving architectures and that adoption of a suitable architecture will speed up the process of drawing-out of concepts and rules. What is a suitable structure in which to store knowledge about design?

The influence of choice of representation on the ease of problem solving is undeniable. Just as the wrong representation can make even the most competent problem solver inefficient (if you don't believe this, try doing multiplication with Roman numerals) a good representation can make a problem seem trivially simple. In addition, a new form of representation can extend the power of a problem-solving strategy into new domains. The program EMYCIN, for example, based on a medical diagnosis program, has been used in other diagnostic domains with some success. Given a good representation, an expert system may be able to 'learn' to do more than the original expert could achieve.

There is more than one way of acquiring knowledge from the expert than by asking direct questions. Methods of inducing rules from examples are being developed by many researchers and rules established by this method have been found to be more efficient in performance to those elicited in the direct way. This approach leads one to speculate that the process of making previously obscure domain knowledge explicit, coherent and logical will deepen and extend our grasp of the domain. It is to be hoped that the domain of design, which has been considered to be so difficult, will benefit from attempts to mechanize it in just that way.

Complete automation of the design process as it has been described here is neither likely to occur in the near future, nor desirable, since the initial specification and the resulting criteria for a successful design are essentially responses to human needs and reflect ever changing human values. However, given an adequate specification and the means to modify it according to human requirements, there is no reason why CAD tools should not be made for all areas of design. In an era of increasingly complex engineering projects, there is every reason why such tools are necessary to technological progress. At the same time, the creation of these tools gives a rare opportunity to psychologists for empirical experiment with models of the complex cognitive processes involved in design.

However, the use of exact models of human performance in the design of CAD tools does not necessarily guarantee a good CAD system. In robotics, the first 'mechanical men' were truly anthropomorphic: in aeronautics, the first flying machines looked like birds. Nowadays, robots perform a subset of tasks which humans perform badly, or find boring, or cannot do at all. Their form reflects their specialized function. Aeroplanes today lack some crucial bird-like characteristics, for example, feathers. Bioengineering solutions are of necessity different to those of mechanical engineering. In a similar way, CAD tools will come to resemble human cognitive processes less, to perform a limited subset of human-like tasks, and will do some things that humans cannot even attempt. Procedures for design which work well when run on general-purpose human wetware may well not be suitable as CAD tools. We all have to start somewhere though: it is to be hoped that psychologists, by accurately describing and analysing human performance, will provide clues to the equivalent of the principle of 'lift' for cognitive science and engineering.

Glossary

The italic entries (eg *decomposition*) are defined elsewhere in the glossary.

abstraction
The application of a simple model to a more complex object in order to: (a) determine a rational *decomposition* into less complex parts, or (b) isolate a particular aspect of function or structure of a complex object.

algorithm
A step-by-step description of a procedure which performs a well-defined function, eg a conventional computer program.

analog circuit
A circuit which handles continuously variable electrical signals (see also *digital*).

artwork
The drawings produced as final documentation for electronic circuits.

complexity
The complexity of an object can be measured by taking the ratio of the number of its parts to their connectedness. An object with a large number of parts, all of which are connected to each other, will be deemed complex by most people.

computer-aided design (system/tools)
A CAD system consists of a set of CAD tools which are usually *algorithms* which automate particular steps in the design process.

constraints
In AI, a constraint is a variable which can be applied in a

search *algorithm* which curtails search in specific directions, eg

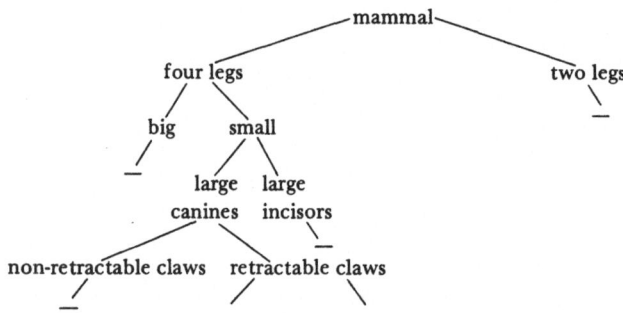

This search through the space of all mammals has been narrowed by three items of information to the much smaller search space of all felines. The assertions that the mammal in question is small, has four legs and retractable claws are all constraints on this particular search space.

criterion

A criterion is a standard for judgement based on a set of *values.*

database

Databases come in all shapes and sizes, but are all highly organized data structures in which search for specific kinds and combinations of data is made easier.

decomposition

Decomposition is an operation performed on a complex object to render it more tractable. Decomposition may be on the basis of function or structure, eg in a car, functional decomposition would split the car into systems (ignition, fuel, cooling, steering, driving, etc), whereas structural decomposition splits it into assemblies (chassis, bodywork, engine, transmission, suspension, half-shafts, etc).

design rules

The *implementation* technology of electronic circuits is only referred to by the designer via the design rules, which are a set of simple Boolean expressions defining limits on the sorts and sizes of shapes in a *layout.*

design tools

A design tool is a piece of software which automates a part of the design process. It may be a component of a *CAD system* or be used alone.

digital
A digital circuit is one in which specified ranges of voltage operate (usually) binary switches, or gates, to embody logical expressions (see also *analog*).

domain
This term is used in the jargon of *expert systems* to refer to a limited area of expert *knowledge* about which a single *knowledge-base* can be built up.

draughtsman
Draughting by hand is a highly skilled craft, demanding a long apprenticeship. In electronics, a draughtsman converts the circuit or *logic diagram* into a *layout* of the actual components or shapes which make up the finished product.

electron beam lithography (EBL)
This is a method of etching patterns on a silicon wafer using an electron beam, which knocks off a layer of molecules from the treated silicon surface. Patterns can be drawn directly on the chip using a moveable electron beam. This process is sometimes called 'ion-milling'.

electronics engineer
An electronics engineer is a member of a design team who deals with the levels of design intermediate between *specification* and *layout.*

expert system
This is a rather overused and vague term which at the very least indicates that the system referred to uses production *rules* as a means of capturing the *knowledge* of a human expert, usually for such purposes as diagnosis, planning or design, which involve reasoning under conditions of uncertainty.

floorplan
A floorplan is a block diagram of the functional *decomposition* of a microprocessor chip.

formal
For present purposes, a formal statement of a problem or its solution is one which can be made in mathematical terms.

The clarity, rigour and precision of a formal statement makes it verifiable. Under this definition, most computer programs are not formal statements (see *procedure*).

heuristic
A heuristic is a rule of thumb used where more well-defined sources of *knowledge* fail because of inadequate or uncertain information. A heuristic program is usually rule-based.

human-computer interface
This is a blanket term which covers all conceivable processes in which a person and a computer can be involved. Here it is sometimes used specifically to refer to a suite of programs which translate the user's input into computer-readable form and *vice versa*.

implementation
An implementation of a design is its instantiation in a tangible implementation technology. Hence a logic design may be implemented as an *integrated circuit* or as a *printed circuit board.*

integrated circuit
An integrated circuit is a circuit whose *implementation* technology is a small chip of treated silicon, rather than an agglomeration of discrete components.

ion implantation
In order to give the silicon *wafer* its distinctive electrical properties, parts of it are 'doped' with ions (usually of boron) which are fired into the wafer in a controlled beam.

knowledge
The kinds of knowledge considered here are: (a) the cognitive resources used by a designer in order to create a design and (b) the *rules* which embody those resources in a *knowledge-base.*

knowledge-base
A knowledge-base is a collection of *rules* concerning a particular *domain* of expertise contained in a highly structured form of *database.*

layout
A layout is the final product of electronics design, being a map or blueprint of the actual geometry of a chip or the

relationship of components on a board.

level (of description/representation/decomposition/abstraction/refinement/ analysis etc)

Wherever the term 'level' is used, the reader must assume that it refers to a partitioning (in the relevant named dimension) of the space of possible designs. Some of the dimensions named above are mentioned both elsewhere in the Glossary and in the main text (especially in Chapter 5). 'Levels of description' are used by Stefik *et al.* (1982) in much the same sense as 'levels of representation' here. 'Level of analysis' is used in the general sense, and implies demarcation of a particular level of abstraction.

logic diagram

The logic diagram is the conventional form of *representation* used by *electronics engineers.* Boolean functions are represented joined in a symbolic circuit which is an *abstraction* of an actual circuit.

mask

A mask tape is the output of a *CAD system* for the production of *integrated circuits.* It is a tape containing a *digital* representation of the patterns which allow different layers to be etched, implanted or deposited on the surface of a *wafer.* The masks themselves are usually made of aluminium on glass produced by photolithography.

printed circuit board

A printed circuit board is a flat plastic card which acts as a mounting for a discrete circuit (one composed of a variety of individual components). On its surface the wires which connect the components are printed in aluminium.

procedure

This is the computer science meaning of the word, equivalent to 'program', and sometimes to *algorithm,* ie the sequence of steps which describe how something is carried out.

refinement

Refinement is best understood as being the opposite of *abstraction:* it is a top-down design process which generates ever finer detail in the design.

representation

A representation of an object is a symbolic model which may

embody in it different kinds of *knowledge* about that object. The following helps to make this clear:

> Consider the following two-person game, which has been called Number Scrabble.
>
> Take nine slips of paper, mark them with the numbers from one to nine, and place them face up on the table. The first player must take any slip of his choice, then the second player must take one, and so on. As soon as either player has any three slips in his possession whose numbers add up to 15, he is the winner. The opportunity to go first should alternate in successive games.
>
> ... most of you will find it difficult to play an expert game. Now look at the figure below:

$$2 \quad 9 \quad 4$$

$$7 \quad 5 \quad 3$$

$$6 \quad 1 \quad 8$$

> Play Number Scrabble again, this time using this figure. The first player should draw an X over each number he selects, and the second player should draw an O. Do you recognise this game? Is it as difficult as you thought Number Scrabble was? What makes it so different?
>
> *Bertram Raphael 'The Thinking Computer: Mind Inside Matter'*
> *W H Freeman 1976*

rule

A (production) rule is a conditional statement of the form
IF (condition) THEN (action)
Its simple form belies the power of rule sets (or *knowledge-bases)* in which individual rules interact and in which quite complex chains of inference can be set up.

specification

A specification is a description of the behaviour required of a machine to be designed. Formal specifications can come close to the ideal of a functional equation, but informal specifications often include details of requirements at a much lower (structural) level.

structured design

A concept made famous by software engineers and computer scientists interested in automatic programming, structured design is a methodology which requires: (a) accurate and complete high-level *specification,* (b) *decomposition* into well-defined modules and (c) recomposition according to composition *rules* (which define legal compositions of

modules). Structured design for electronics engineering was
first evangelized by Carver Mead and Lynn Conway (Mead &
Conway 1980). It is also known as 'top-down design'.

values
Values are needed in order to make judgements: they can be
'rational' (ie easily evaluated, like 'expected monetary value'
in decision analysis), 'irrational', or context dependent. In
the author's studies, values were indicated by a combination
of relative bias of judgements and likely personal interest in a
particular outcome. However, attempts were not made to
accurately measure any of these parameters. The judgements
made by the author are equally affected by her own values.

wafer
A wafer is a thin slice of very pure silicon crystal which can
be treated by etching with chemicals and/or gases, chemical
vapour deposition and *ion implantation* to produce a surface
which displays electrical qualities suitable for making
electronic circuits. *Wafers* are circular, between three and five
inches in diameter, and can be patterned with up to 250
chips. The wafer is sliced into dice, each of which has a circuit
on it.

References

Anceau, F. (1983) CAPRI: a design methodology and a silicon compiler for VLSI circuits specified by algorithms. In *Proc. 3rd Caltech. Conf. on VLSI* Computer Science Press.

Begg, V. (1983a) *Making Cad Tools for Electronic Design More Useable*, Technical Report, MCSG, Department of Computing and Statistics, Brunel University.

Begg, V. (1983b) *Making Computer Aided Design Tools More Useable: A Study of a Complex Task Shared by People and Machines* PhD Thesis, Brunel University.

Bennett, J.S.; Englemore, R.S. (1979) SACON: a knowledge based consultant for structural analysis.
Proc. 7th IJCAI p. 47.

Birtles, P. (1983) Introducing CAE into the aerospace industry.
Electronics & Power 29(1), 59-63.

Bo, K. (1982) Human-computer interaction.
IEEE 'Computer' November.

Bobrow, D.G.; Goldstein, I.P. (1980) Representing design alternatives. In *Proc. AISB-80.*

Broster, D.; South, A. (1983) The role of CAD in semicustom IC design.
Electronics & Power 29(1), 42-47.

Bryant, R. (ed) (1983) *Proc. 3rd Caltech Conf. on VLSI* Computer Science Press.

Carroll, T. (1982) The adventure of getting to know a computer.
IEEE 'Computer' November.

Chen, M.; Mead, C. (1983) A hierarchical simulator based on formal semantics. In *Proc. 3rd Caltech Conf. on VLSI* Computer Science Press.

Claus, V.; Ehrig, H.; Rozenberg, G. (eds) (1979) Graph grammars and their application to computer science and biology.
Lecture Notes in Computer Science no. 7, Springer Verlag, Berlin.

Constantinou, S.; Leonard, R.; Rathmill, K. (1982) Evaluation procedures to be used during the development of CAD systems.
CAD 14, 6 November.

Conway, L. (1981) *The MPC Adventures*
Xerox PARC Technical Memo.

Dahl, O.J. *et al.* (1972) *Structured Programming*
Academic Press, London & New York.

Davis, R.; Buchanan, B. (1977) Meta-level knowledge: overview and applications.
Proc. 5th IJCAI pp. 920-927.

De Kleer, J. *et al.* (1977) *The Explicit Control of Reasoning*
Massachusetts Institute of Technology, Artificial Intelligence Memo no. 427.

Dreyfus, H.L. (1979) *What Computers Can't Do*
Harper & Row, London.

Erman, L.D. *et al.* (1980) The Hearsay-II speech understanding system:
integrating knowledge to resolve uncertainty.
Computing Surveys 12, 213-253.

Fikes, R.E. (1982) A commitment-based framework for describing informal
cooperative work.
Cog. Sci. 6, 331-347

Friedenson, R.A. *et al.* (1982) Designer's workbench: delivery of CAD tools
Proc. 19th EE Design Automation Conf. pp. 15-21.

Gaines, B.R.; Facey, P.V. (1975) Some experience in interactive system
development and application.
Proc. Inst. of E & E Enggs. 63, 894-911.

Guedj, R.A. *et al.* (1979) *Methodology of Interaction*
North Holland Publishing Co, Amsterdam.

Hayes-Roth, B. (1983) *The Blackboard Architecture: A General Framework for
Problem-solving?*
Technical Report, HPP-83-30, Computer Science Department, Stanford
University.

Johnson, S.C. (1982) Hierarchical design validation. In *Proc. Conf. on Advanced
Research in VLSI* Massachusetts Institute of Technology.

Jones, C.B. (1980) *Software Development: A Rigorous Approach*
Prentice Hall, New Jersey.

Latcombe, J-C. (ed) (1978) *Artificial Intelligence and Pattern Recognition in
Computer Aided Design*
North Holland Publishing Co, Amsterdam.

Lenat, D.B. (1983) The nature of heuristics.
Artificial Intelligence March.

Lewin D. (1977) *Computer Aided Design of Digital Systems*
Edward Arnold, London.

Mackworth, A. (1983) On seeing things, again.
Proc. 11th IJCAI 2, 1187-1191.

Maguire, M. An evaluation of published recommendations on the design of man-computer dialogues
Int. J. Man-Machine Studies 16, 237-261.

McDermott, D. (1978) Circuit design as problem solving. In *Artificial Intelligence and Pattern Recognition in Computer Aided Design* North Holland Publishing Co, Amsterdam.

Mead, C.; Conway, L. (1979) *An Introduction to VLSI Systems* Addison Wesley, London.

Milner, R. (1980) A calculus of communicating systems.
Lecture Notes in Computer Science no. 92, Springer Verlag, Berlin.

Mitchell, T.M. (1978) *Version Spaces: An Approach to Concept Learning* PhD Dissertation, Stanford University, December.

Newell, A. (1980) The knowledge level.
AI Magazine Summer.

Newell, M.E.; Fitzpatrick, D.T. (1982) Exploiting structure in integrated circuit design analysis. In *Proc. Conf. on Advanced Research in VLSI* Massachusetts Institute of Technology.

Penfield, P. (ed) (1982)
Proc. Conf. on Advanced Research in VLSI Massachusetts Institute of Technology.

Reboh, R. (1981) *Knowledge Engineering Techniques and Tools for Expert Systems*
Linkoping.

Rem, M. (1981) The VLSI challenge: complexity bridling. In *Proc. VLSI 81 (Edinburgh)* Gray, J.P. (ed), Academic Press, London & New York.

Rem, M. *et al.* (1983) Trace theory and the definition of hierarchical components. In *Proc. 3rd Caltech Conf. on VLSI* Computer Science Press.

Rosenfeld, A. (1974) Multidimensional formal systems. In *Basic Questions in Design Theory* North Holland Publishing Co, Amsterdam.

Rupp, C.R. (1981) Components of a silicon compiler system. In *Proc. VLSI 81 (Edinburgh)* Gray, J.P. (ed), Academic Press, London & New York.

Sacerdoti, E. (1977) *A Structure for Plans and Behaviour* Elsevier Scientific Publishing Co, Amsterdam.

Sata, T.; Warman, E. (eds) (1981) Man-machine communication in CAD/CAM. *Proc. of the IFIP WG5.2-5.3 Working Conf. 1980* North Holland Publishing Co, Amsterdam.

Shrobe, H.E. (1982) The data path generator. In *Proc. Conf. on Advanced Research in VLSI* Massachusetts Institute of Technology.

Simon, H.R. (1973) The structure of ill-structured problems.
Artificial Intelligence 4, 145.

Siskind, J.M. *et al.* (1982) Generating custom high performance VLSI designs from succinct algorithmic descriptions.
Proc. Conf. on Advanced Research in VLSI
Massachusetts Institute of Technology.

Stefik, M. *et al.* (1982) The partitioning of concerns in digital system design. In *Proc. Conf. on Advanced Research in VLSI* Massachusetts Institute of Technology.

Sussman, G.J. (1978) SLICES: At the boundary between analysis and synthesis. In *Artificial Intelligence and Pattern Recognition in Computer Aided Design* North Holland Publishing Co, Amsterdam.

Taylor, R.; Wilson, P. (1982) Process oriented language meets demands of distributed processing.
Electronics 30 November.

Weiss, S. *et al.* (1982) Building expert systems for controlling complex programs.
Proc. AAAI 82.

Index

abstraction, 49-57, 59, 61, 70, 81, 82, 88, 90, 91, 95, 100
algorithm, 53, 59, 60, 67, 69, 77, 91, 95, 100
algorithm library, 95
analog circuits, design of, 17, 21, 23
artificial intelligence, 11, 45, 52, 63, 68, 70, 83, 97, 99
  problem formalization, 45-9
  techniques for electronics engineers, 63-84
automated design, 77
automatic layout routines, 27, 34, 62
  PCBs, 27
  ICs, 29
automatic placement routines, 28
automatic routing, 28, 29
automation, 9, 15, 36, 38, 49, 53, 75, 102

blackboard, 82, 92-4, 100
boundary, of two models, 55
breadboard, 46
BROWSER, 91, 92

CADCONS, 87, 89
category theory, 67, 81
cell libraries, 62
chip design, 22, 23
circuit analysis, 13, 17, 20, 23
circuit design, 25, 50
circuit designers, 72
circuit diagrams, 33
circuit simulators, 13, 23
circuit verification, 24
cognitive engineering, 97
communication, schemes of, 43-4
complexity, 68, 74, 81
complexity (in electronics design), 12, 13, 42, 86
complexity (of electronics systems), 10

complexity control, 12, 22, 32, 48, 60-2, 67, 78, 99
complexity, levels of, 43
component hierarchy, 65
component libraries, 27, 35
composition rules, 65, 66, 72
computer-aided architectural design, 97
computer-aided design
  ideal system, 85-96
  use of, 13-18
  users' needs, 35-42
computer-aided design systems, 11, 13, 15, 16, 24, 25, 27, 32-6, 38-44, 72, 82, 83, 85, 87, 95, 96, 98
computer-aided design systems design, 38, 70, 72, 78, 89, 98
computer-aided design systems, ideal type, 69
computer-aided design programmer, 44
computer-aided design tools, 10, 13, 23, 24, 31, 32, 34-9, 59, 78, 83, 100, 102
computer-aided design, users of, 35
computer-aided engineering, 97
computer-aided manufacture, 31, 97
computer programming, 30
connectivity, 24, 28
consultant, computer as, 77
consultant interface, 92
context principle, 80
context-structured database, 80
control section generator, 74
Conway, 52, 65, 68, 79, see PALLADIO

database, 27, 81
data books, 29, 35
data path generator, 74
debugging aids, 17

decision-making procedures, formal/
    informal, 67
decomposition, 49, 54-6, 69, 73, 85
    complex function, 51
    functional, 64-5
    hierarchy, 77
    methods of, 64
    simple, 51
De Kleer, 72
description levels, 65
DESI, 72
design, 8, 10, 11, 19
    process of, 19-24
design, automation of, 11, 60, 62, 69,
    73, 75
design cycle, 53
design engineers, 24
design hierarchy, 49
design language, 24-6, 73
design library, 73, 80
design methodology, 8-10, 36, 48
design methods, 62
design process, 19
design rule checkers, 29, 30
design rule checking, 18
design rule checking program, 76
design rules, 28, 29, 59, 77
design team, 12, 19, 23, 26, 31-3, 39,
    40, 48, 53, 58, 86, 89, 92, 98, 99
design tools, 8, 24, 33, 60, 62, 85, 92,
    96
designer, 8, 14-6
designer's toolbench, 74
dialogue design, 34, 42
documentation, 36, 58
draughtsman, 25, 28, 37, 39, 40, 42,
    49, 58, 59

ELAS, 77
electronic engineering, 9
electronics, 11
electronics design, 16, 37, 46, 49, 50,
    68, 87, 96, 100
    complexity of, 9-13
electronics designers, 10, 98
electronics engineering, 24, 59
electronics engineers, 14, 25, 38, 40,
    43, 99
electronic systems, 10
EMYCIN, 101
error messages, 34
expertise, 48, 59, 62
expert knowledge, 59
expert systems, 11, 78, 79
extractor, 74

flip-flops, 22
floorplan, 50, 53, 55
formalization, 45, 57, 70
formal knowledge, 57, 70
formal methods, 68
formal specification language, 21, 46,
    48
formal specifications, 21, 23, 52, 53,
    60, 62
full custom design, 22
functional abstraction, 79
functional checking, 24
functional decomposition, 50, 52

gate array, 22
gates, 56
generation, of models, 56
generator, 94
grain size, 56
graphical input, 40
graphics, 29, 41, 80
graphics editor, 39, 86, 92, 94

HEARSAY-II, 82, 93
heuristics, 35, 55, 59, 69, 78, 93, 100
hierarchical decomposition, 59, 67
hierarchical menu structure, 41
hierarchical structure, 52
hierarchy, design, 49
high-level design, 23
high-level design aids, 20
high-level design languages, 36, 85
high-level formal specification, 95
human-computer interface, 10, 15, 16,
    31, 32-7, 42, 43, 63, 64, 77, 78, 81,
    85
    communication through, 33-5

ill-structured knowledge, 98
ill-structured problem, 45, 99
implicit knowledge, 60
inference rules, 81
informal knowledge, 59, 62
integrated circuits, design of, 12, 15,
    17, 23, 29, 30, 59, 87
    layout of, 80
integrated system, 34
intelligent CAD systems
    advantages of, 95-6
interpretation, 55, 56
interpretation rules, 56
initial specification, 90
intuition, 59
IRENE, 74

knowledge, 7
  formal and informal, concepts of,
  57-62
knowledge acquisition system, 91, 92
knowledge base, 91, 96
knowledge, engineering, 68
knowledge level, 70
knowledge sources, 68, 82, 83, 93

labelling (of gates), 26
large-scale integration chips, 7, 10, 63
  circuits, 36
  design, 22
  layout designers, 22
layout, 18, 50, 53, 58, 88
  automatic routines for, 27-31
  logic of, 24-6
  manual, procedures of, 26-7
layout assemblers, 74
layout design, 13, 22, 32
layout designers, 22, 25, 26
layout level, 23
layout verification, 18
logic circuits, 24, 36
logic design, 13, 23-4, 32
logic designers, 23, 24, 26, 27, 41, 58,
  60
logic diagrams, 18, 30, 53, 56, 58, 60
logic level, 23
logic simulation, 17, 24
logic simulators, 13, 24

MacPhitts, 73, 75
manual layout, 26
  procedures for, 26
manual routing, 29
mask making, 31-2
mask tailor, 76
mask tapes, 31
masterslice, 22
mathematical models (of circuits), 13,
  17
McDermott, 76
  see DESI program, 71
Mead, 10, 52, 79, 80
menus, 40
menu systems, 34
meta-level knowledge, 81
multiplexors, 22

nested decomposition, 50
netlist, 24, 28, 60
Newell, 70, 79

object recognition, 50
occam, 73

on-line documentation, 34
open circuit, 41

packaging, 28
packaging (of gates), 26
PALLADIO, 64, 65, 72, 79
parallel models (of design), 61
partial designs, 22, 80, 89, 90
pattern recognition, 41
PIE, 89, 92
predicative calculus, 72
printed circuit board design, 10, 16,
  17, 26, 29, 46, 87, 96
  designers of, 22, 27
problem solving, 70
production rules, 100
programmable logic arrays, 22, 74
program transparency, 78
propagation of constraints, 73

random logic, 10
recomposition, 95
refinement, 49, 54, 55, 56, 58, 61, 69,
  88
register transfer language, 92
Rem, Martin, 69
representation (of a design), 11, 17,
  26, 47, 49, 52, 53, 57, 59, 67, 69,
  73, 79, 81, 82, 86, 88, 90, 94, 95,
  99, 100
representation (of CAD tools), 83
representation hierarchy, 52, 54
routing algorithms, 29
rule bases, 98-9
rules, 50
rules (of interpretation), 61

semiconductor design (complexity in),
  14, 22
semiconductor industry, 36
semiconductors, 8
sequential models (of design), 57
set theory, 67
silicon compilers, 30, 48, 52, 66, 73,
  74, 87, 94, 100
Simon, Herbert, 45, 48
simulation, 37, 40, 80, 86, 89, 94, 95
simulator, 75, 95
Siskind, 75
SLICES program, 72, 79
Smalltalk 'browser', 80
software, 23, 24, 62, 80
software design, 10, 31, 34
software designers, 98
software designing (with CAD tools),
  32

software engineer, 78
software engineering, 52, 62
specification, 69, 70, 72, 79
specification aids, 101
specification language, 89, 92
specifications (for designs), 20, 24, 47,
    52, 56, 59, 60, 61, 88, 89, 90, 92,
    95, 100, 102
specifications, of an indexed partial
    solution, 71
SPICE, 23, 31
standard cell libraries, 36
standard cells, 22, 74, 96
Stefik, 65, 79, see PALLADIO
sticks diagram, 30
structural decomposition, 50
structure (of a design), 10
structured design, 23, 37, 81, 100
structured programming, 11, 52
Sussman, 73, 76, 79
    see SLICES program
system level, 23
system specification, 21
systems analyst, 47
systems complexity, 10
systems designer, 21, 40, 59, 99

task complexity, 40, 99
task model, 44

technical knowledge, 48, 59
technology independence, 48
testability, 24, 32
transformation (between models), 56
transistors, 9, 22
translation, levels of, 43
tree structure, 49

uncommitted logic array, 22
use, of CAD for design, 16
user model, 16, 34, 38, 42, 86
user profile, 38, 39

validation of models, 56
verification, 40, 75
video games, 41
very-large-scale integration, 52, 63, 64,
    73, 99
    applications, 96
    chips, 53
    design, 23, 36
    hardware, 23

wafer, 9
wetware, 87, 102
work-station, 16, 30